STEPPING ASIDE

(Via the Sixth Sense Door)

The Key to Insight

By U. Notmi

www.ReallySo.org

i

ii

CONTENT

INTRODUCTION

Synopsis

If you are serious about the insight practice then you have arrived at the right place. This is the only place that presents a detailed explanation of how the mind forms, and how the mindfulness practice affects it. The practice is natural, intending to overcome the biases of the mind, hence, to begin with, it would be helpful to know exactly what these biases are. You will find it here, as well as an integration with science and real life, which is essential in order to attain the required conviction.

With all due respect, the Buddhas (wise men) do not have the means to explain (connect with science) nor the motivation to detail. Scientists, on the other hand, tend to complicate and ignore this insight. Evidently, thousands of years have passed, and this topic has remained unclear. Here it is different. Based on a scientific model (The Foundations of the Mind), all the ancient sayings are combined into a coherent structure, so you can know – it is Really So!

The Buddha cannot be comprehended directly, but rather indirectly, because there are inherent flaws in our assumptions. Thus, the ancient practice is indirect, and so should be its explanation. Accordingly, the book begins by presenting briefly the physics of the mind (a general model) and then uses it to clarify the ancient instruction. It makes the instructions clear, and solves the main problem – that intuition is too difficult to communicate, because it is not tangible, and each word can be interpreted in many ways.

For proper practice all you need is to set your mind in the right direction, but what exactly is it? Who can point it out for you? You must reach it by yourself, and therefore, you may need the details presented here. The ancient instructions state clearly that you do not need to read too much, since the more you read the less focused you are and the less benefit you gain. Often just a few sentences that put your mind in the proper direction, will do much better than many books. Hence, it is presented here concisely, trying to be as precise as possible, focusing on achieving the aim – a true, simple and direct understanding.

The scientific explanation presented here shows how all meanings get assembled to form the reality as we see it (our mind's content). With this knowledge, you will know how to dismantle it back to the original structure, which is Buddhahood. It is made simply by maintaining the proper mental orientation, which is the main thing you need to find. Once you have attained it you are onboard, and the "boat" will take you to the promised land.

The truth is when everything fits together and this is the only thing you need to verify. This may sound trivial, yet, the normal approach is that everything must fit the framework [of your mind], which remains intact, while therein lay the flaws. When everything completely fits, the framework is done with, and you have attained Buddhahood. It is about the release of fixations. Not easy, yet simple.

Preface

Stepping Aside is based on the Foundations the Mind, which at its core is presenting the actual physical (neurobiological) process that underlies learning, perception, and behavior. It continues by showing the implications of such a process, what biases of perception must be, the damage they cause, and the way to be released from them. That's why its original title was The Path to the Boat – it clarifies what this boat is, how to get there and why.

Similarly, Stepping Aside can be seen as the "key to the boat". If you haven't arrived at the boat then you don't need a key to enter, as it will be of no help. This book was made for the serious ones, those who have already made their verification, concluded that this path is correct, and deeply feel that they need that change. Not just like, interested or wish, but rather a real gut feeling of need with some urgency, because if this path is correct then our major conventional beliefs are incorrect, and they misdirect and embody a heavy price. Without such conviction, the pitfalls cannot be bypassed and the boat cannot be reached.

In this book, I repeat briefly the major arguments and conclusions in order to provide a complete picture, in case you didn't read or don't remember the Foundations of the Mind. However, the main objective is to make it actionable – to help you step onboard and get going. For this aim, it is required to concentrate only on that which may bring results, and leave aside other perspectives and details. It is not intended to tell you something entirely new that was not mentioned earlier – it is only a different perspective, but one which makes a difference.

The boat is the Buddhist path, the practices of which are detailed in many books. I have already presented the major ones, and do not claim to innovate in this regard, yet my emphasis is different in two respects:

1) A focus has been made on the error – saying and demonstrating it whenever it is there. Not to hide it or be politically correct. It means that the detailed and concrete explanation is taken seriously here. Similarly, regarding the practice, one needs to see the error directly, mainly about oneself, by oneself.

2) Really So, meaning, that you already have the relevant knowledge and do not need to search anywhere, but rather it is here, now, and all the time. It implies that while normally there are defined hours for strict practice, and in between, you need to watch yourself in action, I would make that "in-between" as the main object. Because it is Really So, hence, there is no need for any artificial means. Just realizing these two words already puts you onboard. So, you can simply watch yourself preparing a cup of tea, walking to the shop or speaking, etc., and even birds can teach you something if you care to observe.

The first thing to remember when dealing with this subject is that science has failed to explain how things are represented in our brains, how they are there in memory, and how they are learned and recalled. This relates to all opinions and perceptions, including the whole world that you constantly see around you. All things were learned means that all things must embody bias, and the perception that they are an objective reality is the primary one. Thus, we have something to discover here and a reason to continue in this investigation. The objective orientation we live by is just practical, not absolute truth, and it is practical only in a limited range (to solve basic needs and communication), beyond which it may be destructive. Hence, unlike what many might think, attaining the insight may also be very practical – it saves you from much trouble and needless effort.

It is almost trivial that every issue requires a different means to be solved, and when it comes to the foundations, insight is the means. You may need an airplane to return to your city, then a taxi to get home, and insight to be at home there, and anywhere. You wouldn't try to get to your home with an airplane, and similarly, you shouldn't try to solve fundamental issues with false objectivity. Mindfulness is the way for that.

While the focus here is to help the truth seekers to get onboard the "boat", the mental processes presented are the same for everyone. It is important to remember this point due to the huge distortion and distraction we are facing anywhere we look. This is similar to the support the Buddha advises us to get from the Sangha (friends taking a similar path), but focusing on the error (to gain support) instead of a truthful example.

It also clarifies that you are alone on this path, and not realizing this truth is a major hindrance. This is an important point. Science, too, has failed in this regard, and therefore they ask the wrong question, one which obviously cannot be answered. Unfortunately, most, if not all, geopolitical and other turmoil have their roots in seeing the internal as external, refusing to accept the ultimate truth, which is — each mind is alone, constituting the whole world. May we all overcome these biases and falsehoods.

The Mechanics and the Resulting Biases

The "physics of the mind" refers primarily to the representation method of the brain, assuming that it is consistent and that it applies to all aspects of mental activity. It will be summarized here briefly as a basis for the understanding of the practice, which is our main subject. The technical theory may give the essential conviction that otherwise is so missing, and without which, there cannot be much progress in this difficult and tricky path. It also enables more precise directions and a much clearer way to communicate them.

The term "representation" refers to the connection between the physical aspects (e.g. neurons) and the mental ones (e.g. recall), and since they seem so different, all we can and need to do is to show consistent correlations and compatibility. Correlations are what representation is all about. For example, a photo (representation) is never similar to the real object that is seen in the picture. It only correlates in some dimensions, while others are completely different. Similarly, the mind (which includes that photo) enables good functioning due to good correlations. However, there are also other dimensions, which are biased, and that is normally repressed and ignored.

Assuming that the solution presented here is correct, psychology becomes very simple and clear, if not outright obvious – all its aspects emerge in the brain according to the same principles (the representation method), which thereby constitute its real paradigm.

With that said, let's summarize the main principles, beginning from the beginning.

We can view all meanings in the mind as made of various combinations of features. An object is a summation of features, feeling is a feature, classification is made according to features, and so on. Words, too, are mental objects, made up as summations of features, where the grammar is simply part of their features, defining their role in the sentence and where they may fit, similar to the way their other features define their meaning in other respects. The features of grammar also constitute the classification "grammar" as well as its sub-classifications, and this too is similar to other features (being basis for classifications).

All features are manifestations of neural projections, where the meaning arises from the source and the destination of each projection. It is a circular definition, and it is fractal. The structures in which the neurons are organized (in the brain) also have a significant role in determining meanings, and these, too, are fractal, and will serve the main discussion below.

The "mechanics" (of the brain) that make all meanings and functioning are basically simple and consistent, and can be described as "task management", which is intuitive and easy to grasp regarding activity – each task breaks down into subtasks and serves as a framework to its subtasks. Each

9

subtask needs to be accepted by the task above it (its framework) in order to be completed. In this manner, each level controls and directs the levels below it. This is realized as a three-level circular structure in which the higher level (task) projects into one level below its subtask. It is a fractal structure, meaning that the same principles apply to all levels, and in all dimensions. The subtasks are being summed up to the task above (when completed), which thus is allowed to be summed up to the level above it (to be completed too), and so on, until the super-task is completed. This hierarchy constitutes a verification that a course taken can be completed. The verification is inherent in the structure; it is made before the first step is taken and continuously thereafter, and it is an essential element for success and survival.

This representation structure can also be seen as frameworks and central projections, and that clarifies how the same essential structure that manages the activity (task management) also generates the meanings (mental objects) that are involved in the activity management. The summation of features forms in many levels (of "tasks" or frameworks), where an observed object is the top one, and each feature is simply a projection of neurons, which gets its meaning from its origin and destination. The observed features, too, are built up by many levels, of which the basic ones cannot be described by words. At the higher levels, it is easy to realize how features are associated with one another or attributed to objects. Following such an association or attribution the result appears to be an objective reality, although it could have formed differently.

Any representation must involve biases. The main biases, which contribute to the understanding of the insight practice, are mentioned below. After all, the practice is there to attain the truth, which is to overcome these biases.

Constancy and separation – An object is a summation of features in focus, and as the summation changes, so does the object. Each feature can get into focus, and thereby it becomes an object. This means that in contrast to the way things seem, they are not substantial and they lack inherent meaning (no-self). Since each act of focusing creates an object, we can only relate to objects, and when a feature is in focus (becomes a mental object), it is not the same one that served as a framework, a basis for classification or as an element in an object. Since all things are summations of features, they extremely overlap, in contrast to the way objects normally appear in the mind.

Only the object – Characteristic to the attribution process is that it relates only to one object while there may be others with overlapping meanings, and therefore, they should be subject to the same attribution. This marks the difference between knowing and understanding, hence overcoming this bias is a major concern here. In this discussion, it is done by repeating the same ideas while explaining different terms and practices, showing them as different perspectives within the

11

same fractal structure, and inherently overlap. As I showed previously that information processing, learning, perception, recall, and memory are the same thing (only different terms and perspectives for different usages), so I do here regarding higher levels of that same fractal structure.

Entire object – Attribution is always to an entire object, while most of its features are not relevant for that matter. That creates an inherent bias. It can be argued that the object itself was formed (in the mind) due to the attributions (relating) to "it", and without them, it would not exist (as an object). That is, an object is simply the highest level of the summation of features for the specific matter (usage). The attribution structure implies viewing the object "from the outside", as a compact and unified entity, in contrast to the way it forms [in the mind]. Again, it is an inherent bias, which is the basis for the perception of objectivity. It is the reason why the three characteristics of existence (as defined by the Buddha) are not seen, and knowing this may help to overcome ignorance (see the characteristics) and attain this insight.

The framework – "The" refers to the top level of the fractal structure, which is made of many levels of essentially similar structures. It defines the border between you (the "I") and the external reality, what fits for you, what you must do or avoid, etc., and by that it actually creates you. While it seems that you act according to your free will, the will (framework) controls you, not vice versa. Activeness is always under a framework (will), hence it is never really free, and more precisely, it is stemming from cause (the will), which means passiveness, like everything else. This is an example of the bias inherent in attribution – "my will" is an attribution between two mental objects (me and the will) that one of them is dominant (me), while the truth is that "will" is not an object and does not belong to me, but rather it controls and makes me (framework). It is also an example of how a feature (will) can become an object in an instant (when in focus), and that frameworks are made of features. Furthermore, it shows that all definitions and aspects of the mind extensively overlap.

Objectivity – Our deep belief in objectivity is the origin of endless mistakes, since not only that objectivity is implausible, but also the belief in it prevents a true verification. If you are objective then surely you see correct, and all confusions are attributed to external entities, thus your flaws will persist for life, and mislead others as well. At the most fundamental level, which is the theme here, it concerns the general structure of objects and attributions between objects. This presupposition is taken for granted since all things are represented in this manner. Everything that does not conform to this assumption is repressed and denied on the spot, and does not arise in consciousness. We are utterly deaf to anything that does not fit our basic assumptions. Yet, this structure arises from memory, together with specific meanings (of the objects of reality), and it is not at all objective.

This blind belief in "reality" is the essence of ignorance. It is evident regarding our "value categories", which evidence and reason are ordinarily forced to conform to them [20, p. 25]. Yet, on our way to the insight, we must realize that it is much deeper than just a few categories, and unfortunately, some values that we hold dear, completely block true understanding. Please note the inherent contradiction between free will and objectivity (both we deeply believe in) – If evidence and reason are subject to wills then our seeing is a wishful seeing – not at all objective. Eventually we will have to admit that perceptions that form by brains cannot be objective. See the appendix for a shocking demonstration.

In this regard, it may be revealing to mention a major Buddhist concept called "clear comprehension" (Sampajañña), which has four main aspects. Its second aspect, which is relevant here, is called suitability (Sappaya); that is, one is required to consider whether the intended action's timing, place, and personal capacity are appropriate [Wikipedia]. Hence, viewing the normal state as ignorance (lack of capacity), Buddhist scholars normally refrain from explaining the truth to laymen [see 22]. It is important to keep this point in mind since we are trying to reveal aspects that others didn't solve. For that aim, it is essential to clarify where they missed, both in order to verify the rationale and avoid the pitfalls. When both, the outlook of the listeners is unfit, and the motivation of the teachers is lacking, the truth remains accessible only to very few, which is the problem this book tries to fix.

The "suitability" concept implies attributing less authority to self and others, and such an attitude is vital in our mission. Yet, at the same time, the approach taken here violates the principle of suitability by "saying too much". Naturally, we need to adjust some rules to achieve a new result, which then may not suit the unprepared ones.

The direction of sight   Before detailing how things of the mind emerge in the brain, it is important to take seriously the biases presented above, and particularly that of objectivity. It is detailed in the Foundations of the Mind, and illustrated under the title "direction of sight". You need only a few seconds to get the main point, yet, probably months or years to understand it. Hence, we are here to clarify this topic and elaborate on the practice. We are speaking here about a paradigm shift regarding assumptions that prevail also while reading this volume; therefore, it might be a good idea to take a look at the "final notes" at the end, before delving into details.

FORMATION

<u>Following Authority</u>

After this introduction we can try to get onboard, assuming that we have already checked the detailed analysis, concluded that it is "really so", fits for us, and that there is no way to escape this truth. Still, before getting into the boat, a few more words to clarify why things have remained obscure until now, despite the efforts made by so many intelligent scholars. This boat might be unfamiliar, yet airplanes, we all have used already. How could you have the courage to enter into this tube, which rises so high and it is not clear how it finds its destination? The answer is simple, everyone is doing it, so it must be OK. The authority of others, of organizations, experts, and the like, is the driving force, and not only concerning flights. As a rule, views are accepted based on authority, since in most cases what alternative is there? It relates to practical issues (such as flights), as well as opinions and beliefs, which after being accepted are a part of you. For this reason, it is very difficult to present a solution that is not in accord with prevailing assumptions, particularly with regard to the top-level (I, framework, reality). Such a solution is simply not seen, and even when presented, no one pays attention to it. They may claim that it is missing and needed, yet it must fit the classification they expect in order to gain their attention. Everything is dictated by the framework, that actually blocks anything that may conflict with it. This is functional, and thus may serve an important role, yet it involves an inherent insincerity. We tend to say (and believe in) what seems to fit expectation, even if imprecise or untrue, so long as it serves the immediate needs.

So, also psychological research is characterized by treading the beaten tracks and dependency on prior classifications. The great psychological discoveries are simply the willingness to go against the current, where the knowledge and insights are not really new, and it didn't require a special intellectual effort to uncover them [see 6]. It was required to conduct an "objective" scientific experiment, and present it in the respectable channels (by "scientists"). The findings themselves, too, for the most part, show an excessive tendency to conform, repress contradictions, and be dependent upon others' views, presented in various ways [6]. They also show that reality and imagination are strongly combined, which fits the conformity aspect as a means to conclude what reality is.

Following the norm necessarily implies the repression of that which is different. Naturally, psychology relates to repression and denial, but only of the "subject" (being investigated or treated), stating it from an "objective" standpoint, which is the greatest denial of all. Above all, the contradiction between their views and the law of cause and effect is repressed (and denied). By that, they violate the foundation of science. This repression stems from a cause, of course, it is a natural result of the mental process, which in effect, lacks any activeness or objectivity. The belief in objectivity is simply a manifestation of this repression.

The Gestalt psychology seems to be closer to the truth in this regard (of evolving processes), but it is about "something else", regarding which it is acceptable to claim that it stems from causes, while about us there is a psychology of personality, which already occupies that space. Why the various psychological theories are not integrated is a fundamental question that demands an answer. The reason is that their advocators have interests and needs, which are not the pure truth, and objectivity is non-existent.

They need to get for themselves, and for that aim, theories need to be complex, impressive and numerous, and registered to their names. Furthermore, psychological theories need to be accepted, that is, to fit the norm, and here there is a problem – they are about ourselves, and not everything is allowed in this regard. Yet, based on the law of cause and effect, we should expect that all forms and aspects will form in the mind according to the brain's features, which means that the same principles apply to all, including me and my friends and all events, and this must not be said. Hence, the Gestalt focuses only on some forms.

A professor, like anyone, does and says what seems to fit at that moment, and so every moment. The definitions of "what fits for me" are the framework, and they include the demand for objectivity. The scientists do not explain simply because it doesn't fit them. Whether consciously or unconsciously, this demand affects scientific theories, and it certainly doesn't permit "beginning from the beginning", which is what is required in this case. They are professors already and having worked so hard to gain that status, no way there is they will consider beginning from the beginning. Nevertheless, the lack of an answer is evident. We see but refuse to believe, due to the heavy authority of science, and the necessity to follow the beaten track to survive. However, true understanding requires us to begin from the beginning, so this is what we shall do.

The Center of the World

The perception that I am the center of the world is a source of endless failings in life, as it is in science. It overlaps the belief in objectivity, looking outward, and therefore I am the center from which everything is seen. I see outward means that the seen are real objects, hence others must see the same objects. Similarly, the basic attitude in life that I am being seen (the framework) implies that I am the center. It creates absurd situations. For example, your friend is convinced that he/she is the center of the world while obviously, you are the center. The existence of a few centers to the same world is utterly impossible, hence, your friend's world cannot be the same one as yours, even if everyone will claim and demand so. Yet, although the objective perception is completely refuted, the framework represses this contradiction.

Logical explanations do not change this intuitive perception (that I am the center and objective), and the illusion is maintained.

This orientation as a center embodies the belief in one's own opinions. Simply put, I am the measure of what is correct, and therefore, whoever thinks differently is wrong. It is a source of many confrontations, because obviously, any other mind has a different center, includes different perception and a different reality. Eventually, the incompatibilities are disclosed, hence the confrontations. Such inadequacies pertain also to scientific discussions, and the bias is at an extreme when we (the mind) are the subject.

Since I feel that I am the center, active and real, also what "I see" is real and objective, and the truth (of becoming) is denied, while naturally, it should be the main subject for science. For example, "I am conscious of the sensation in my leg" – it includes an object "sensation" within a larger object (leg), an unclear action (conscious) and an object "I" that performs it. What exactly is this activity of being conscious? The "I" that seems to perform it is not actually its result? Then where is it present? And the sensation itself was there before the consciousness or is it simply [an element of] that consciousness? These are attributions between objects, which are biased results, yet seem to be objective reality. The bias persists due to repression and the denial of the formation process. This brings us to the sixth sense door, which may help us to recognize and overcome the biases, and eventually attain the insight.

The Sixth Sense

To the familiar five senses, the Buddha added one more, which at first seems very different, yet a closer observation reveals that the principles are the same. The objects of the world are perceived via the sense doors (eyes, ears, ..) and form in the mind in a process that, according to the Buddha, involves a structure of nama and rupa for each sense. This distinction between nama and rupa means that everything forms, and therefore is internal. However, the Buddha left this point obscure. Nonetheless, the universal law that there are no results without causes is kept, as required in every scientific discussion. The same fundamental structure of nama and rupa was applied by the Buddha to the way the body is controlled and activated, and this is a fractal definition. Accordingly, the mind is that which sees and controls the body, hence, in our terminology, the mind is the framework. There is a body (rupa) and framework (nama), and when joined, they form the perception of a unified compact "I" (a summation of body and mind). The framework is the constant assessment "what fits for me", that is, where and how the body should be, and what it has to do. Hence, the practice instructions to see only the body position, without an "I", also means without the framework. As already stated, this is also the destination – enlightenment is simply the prevailing perception after the framework has been eradicated (detailed in The Foundations of the Mind).

The fractal principle is now clear – the framework always notices and activates the level below it (via feedback circuits; the three-level structure). So it is concerning all levels and all sense doors (nama for each rupa), including the mind (the sixth) sense door.

Normally we are not aware of this distinction, and see unified objects, and this also refers to the object "I". I feel active because, although the body is activated by the framework, they seem unified (as a compact "I"), and thus the activating force is not external, but rather part of me (my will). This clarifies that "I" and "framework" are overlapping concepts, though while relating to them they are separate mental objects, which is a normal representation bias.

The awareness of the mind sense door (higher level) is a main subject here because it constitutes the weakening of the framework, and by that, it facilitates the awareness of the lower levels (the normal five sense doors), to which most of the published instructions are directed.

Formation of Meaning

The learning of what are the things in the world (including oneself) is the formation of their meaning in the brain and mind. This gradual formation is revealed when watching infants, that still have much to learn, hence it is evident that they do not simply see things as they are outside. After this learning has occurred, most of the information resides in memory (brain) without awareness of its existence, and it emerges in the mind in dependence on circumstances (sensory input and thoughts). Learning and recall are not two separate processes, but rather, upon exposure to a new combination of inputs, new routes (connections) are formed in the brain, and their activity is the change in perception, whether these routes became at the same instant (learning and perception) or at a prior event (recall). This concerns all things and means that at this moment (and always) all meanings arise from memory, even when sensory input is required to activate them (to generate recall).

For example, distinguishing between the internal and external is part of the initial basic learning of a baby (that one had been). Concurrently, the "I" and "what fits for me" are defined. This separation between me and reality, and the definition of the possible and desirable interactions for me, are the framework. My features were learned and attributed to the "I" during the activity in the same way that all other objects gained their features (passively), as different from "I learned" (active; attribution). The activeness we believe in is one of the attributed features. It doesn't fit the laws of nature and relates only to the "I" while regarding all other things we remember that they form by causes. Hence, we must be wrong regarding ourselves.

The perception of space, too, was learned. For instance, where sound arrives from is a sensation in the "ear base", which was established with the aid of visual input and the correlation between these two inputs. Now it is very clear what is tea, what is hot, how it affects me, what taste it has, that it may spill and in which way, how to hold the cup, ... but this is all memory and therefore not at all self-evident (not external nor objective). After I have completed my tea, I paid and returned to my residence. What is required of me and the waiter is memory, where is my residence and what is a room are also memory. What is a key, how to use it, what to expect – all these meanings arise from memory. Even how to move my hand is a recall.

Everything is represented as neural routes, which were formed according to acceptances in the various dimensions. Thus, every object can be viewed as an aspect of acceptance (of its features in a framework), and the same applies to the features, which too were formed in many levels (of acceptances). For example, the association between sound and sight as part of the object's definition (as making the sound) is essentially acceptance, and this applies also regarding the way oneself was formed (in memory).

Acceptance can be seen as "end-of-task", whether the "task" is part of the interpretation of reality (the formation of an object) or an actual task to be done, and whether it is part of the initial stages of the interpretation or of the high levels of attributing a feature to an existing object. Thus, every attribution is an "end of task" (at that level), and when the task is a real activity (conscious attribution; event) it also means a transition from present to past and the creation of the primary time axis.

An action is attributed to the "I" (under the top framework), and then it is "mine" (I am a framework of my actions). For example, while riding my bike, an identification of a new restaurant was formed, followed by an attribution "I have seen", but the attribution was first, and "I" formed (in the mind) due to that attribution (and the preceding identification). While sitting on the balcony, the hand moved toward the glass of water, brought it up, the mouth opened, and the attribution has been formed – I drink. Again a confusion between the cause and the result. The lack of activeness is less clear when "I" have a part in directing the activity (as opposed to incidental identification), yet, "I" is a mental object, a representation, not something that can do, but rather it is part of the choice-making process, and since it is the most adequate available object, it gains the attribution (an aspect of acceptance in this level).

Everything is learned based on the same principles – what are the objects in reality, what I am, what fits for me, activeness, objectivity, and so on, yet, not everything that is learned is accurate. There are inherent biases that stem from initially flawed interpretation (dependency on order of exposure), and this is why we are here – trying to fix the distortions. This is, of course, the approach of the Buddha – that our perception is not objective and that a different way of seeing can form.

As part of his teaching, the Buddha presented the "wheel of becoming", which too often is misunderstood, probably because of its terms. In this cycle (wheel), "birth" appears after identification and "clinging", and "death" comes right after birth. In other terms, an attribution is "birth", and it constitutes the "end-of-task" ("death"), and then remains only the framework, which is "ignorance" and the first stage of a new cycle [see 21 for a detailed discussion]. Ignorance is the lack of awareness of the formation process, and it includes the belief in objectivity and activeness. Most of us may find it difficult to relinquish this belief and the power it entails (over others that believe), but without the misconception of activeness, everything is amazing – the becoming itself is, in every instant, intricately innumerable miracles.

Activeness

The difference between action and activeness is the apparent free will in the latter. Yet, you don't choose your will, hence, there is no free choice, and freedom remains just a feeling, though a favorable one. Activeness feels nice because it embodies acceptance in the framework, which is the goal, and hence it also constitutes the lack of real freedom. The action is made within the "task management" structure, in which each level needs to fit the one above it (be accepted); hence, this is illusory freedom. In passiveness, on the other hand, there is no acceptance, consequently, it is normally less regarded (incompatible to the framework); yet, it is the way to attain freedom (from the framework). Evidently, there is great confusion here, manifested in the inversion between the cause and the effect – "my will" means that I am the cause, while the truth is that I am the result of "my" will and other possessions and classifications, which are wired in "my" brain.

Due to the assumption of activeness, it may sound strange that it is possible to act properly without conscious attention. As the saying goes, when the unreal seems real, the real seems impossible. To demonstrate this point, think of a fly or a fish, which functions very well, yet what kind of consciousness do they have? The simple truth is that although our familiar views seem so real, we have no idea what is consciousness, how it forms, and how it can be active. The distinction between "I" (active) and "consciousness" (passive) hides this contradiction. This is a manifestation of the bias of attribution solely to the object, yet the contradiction remains (the two overlap), which means error.

The task management structure presented here solves this contradiction and shows how all the familiar activities can be done without anyone doing them. For example, I walk back home (framework), the legs are being accepted by the ground (a framework for them), my thoughts while walking are generated in order to form acceptance within another framework, and their internal structure is of many frameworks and objects. Recognizing the shop around the corner defines another framework (destination), and when the cellphone rings, it becomes a framework for the hand that reaches it. All actions are generated by a given dynamics, while "I do" is just the belief in the story – in the attributions that were made, and in the framework. The destination is always a framework and always imaginary, even if functional. The insight practice is designed to escape the framework, so that the task management process will continue "without me" (no-self).

Example: Upon finishing the writing of a sentence, the hand removed the reading glasses and put them on the table. The glass of tea came into attention. The hand reached out, grasped the glass, brought it up, and the attribution "I drank" was formed. It "fits me" to drink, and this is the story and the belief in it (and in the framework that controls it). If it is not

"I" that is doing, then the framework is contradicted – insight is inherent in passiveness.

The Ultimate Truth

The misconception regarding death is in the perception that all things will remain, and only I will not be here anymore. Death is misunderstood because reality is not understood, hence the understanding of death is a means to understand reality – what and where is it. This linking may not be clear due to the bias of attribution solely to the object, yet when it is clear that in reality everything forms, is present for a brief moment, and thereafter fades away, then it is not so different from death and rebirth. The truth is that all mental things come from nowhere and go to nowhere, and everything which is not mental is undefined, un-relatable and unperceivable. Simply put, your reality came to be after your birth, and before that, it was not defined. Another person, in another place, knows other things and lives in a different reality. Then what is it? The awareness that there is an end, and what its meaning is, may stir up the observation and clarify its urgency. The true understanding is without words, without explaining to others or asking others, but rather directly, about yourself, here and now.

OVERCOMING THE BIASES

Inward

All the definitions and explanations concerning the mental process, important and interesting as they may be, cannot be a substitution for the direct understanding (insight), which is without words (much as "green" is inexplicable). Attaining the Buddha's insight requires the observation inward, instead of looking for attainments in the external world. Accordingly, before we analyze and evaluate the various techniques, and choose the best one for us, we should elucidate first what is this "inward", which apparently is so clear but actually forms a confusion that hinders progress. "Inward" does not refer to sensory content or to the content of thoughts, whatever it may be. It also doesn't refer to gaze into the body or brain. So, what does it refer to? The mind is just concepts, mental objects, features, and categories, so the "in" could mean within a certain framework and not others, and that doesn't help anything. Thus, the intended "inward" is to observe how things form in consciousness. It relates to the way mental content arises from memory (recall), including sensory perception, which normally is regarded as "outward". After the initial contact (in the eye, ear, etc.), the sensory input is going through complex processes in the brain. Such processes are essentially a recall, and they can be generally discerned, (i.e. realizing nama-rupa), which is "inward".

The focus here is on the sixth sense door, which is undoubtedly inward – to be aware of how I, my body and the objects in "my" mind change from one moment to the next, as "task management". It concerns the familiar objects and actions, but from a different angle – from the side – without involvement in the story, which thereby, its objects are seen as forming (rather than existing externally). Such distancing is from the framework, from its directions and its acceptance. Just to observe whatever arises as it is. Inward equals alone.

Example: Drinking from the tap. The tap gets closer to the face, although it seems that the face moves to the tap. It seems this way because reality is a fixation of our minds (gaze outward). Actually, every motion is relative and all motions are represented in the brain and present in the mind, and nowhere else (and the awareness to that is "inward", and what is really outside will always remain undefined). Now I sit in the garden. The garden and the world are a fixed construct, and I am in them, but the previous drinking more clearly arises from memory. "I" also arise from memory, so what is "I"? The "I" is the summation of features that have been accepted. Accepted in what? In internal frameworks. It renders all choices untrue, even those that were benefiting. When it is understood this way, the direction is inward.

General Emphasis

The original practice, as presented by the Buddha, remains, yet the physics of the mind enables more detailed directions and somewhat different emphasis. First, the overlap between the various aspects is clearer now. These are not different things; it is not a list that one needs to follow. Rather, these are different ways to express the same idea and direct to the same mental state, which facilitates the desirable change. Second, the practice doesn't seem strange, mystical or unrealistic anymore. The insight is a change in the brain's connectivity, it forms naturally by the normal dynamics, and it happens gradually due to a change in the structure of the neural activity. A change in mental orientation is a manifestation of a change in the brain's activity (the combination of the active neurons). All the instructions for practice here and elsewhere are intended to form this activity structure (mental and neural), and based on the Foundations of the Mind it is possible to elaborate on it and communicate it much better. Otherwise, intuition is very difficult to communicate, as is evident throughout Buddhist history.

Here the emphasis is on the sixth sense door, enabled by detailed technical explanations. Simply, it is easier to release clinging in the higher levels first, because when these frameworks prevail, they prevent mindfulness to the lower levels (five sense doors). It is the "stepping aside" that matters, aside from the framework, yet, when it is too strong, this "stepping" does not occur. Thus, the demonstration of the error at the higher levels, as is done here, facilitates the initial progress (where it is easier), and by that further progress is enabled, similarly to solving a crossword puzzle.

This is consistent with the Buddha's instructions for the thoughtful people, as opposed to the lustful ones, which should focus on the body [18]. Analytical thoughtful people may get lost while observing the body [19]. Unfortunately, this may mean that most of the instructions today are misdirecting for most people. "Thoughtful" means under a framework, which prevents proper concentration on the body.

Accordingly, an emphasis is made on mindfulness during daily activity, as opposed to many hours of quiet observation. Again, it is the understanding of the physics of the mind, which makes this more effective and valuable. Quiet observation, too, has its role, and naturally, in each stage on the way a different practice may serve better. This brings me back to the first emphasis, which is the understanding of the overlap in the way the various methods operate (affect). Needless to say, all methods involve a passive orientation in order to avoid a mental trap, which blocks progress. For example, quiet observation in a certain position may seem like a task that one needs to fulfill, but if so, then the framework prevails, and the practice is not a real one.

Really so   These two words were essential for the explanation of the physics of the mind in the face of the all-inclusive denial of the fundamental truth. It means using the available knowledge and build up without delay, and this requires conviction ("really so"). Now, when the theme is the actual progress on the path, the internal conviction is crucial. While the general confidence in your capabilities must be maintained at all times, it is important to remember that the way you see yourself is incorrect (as a center, active and being accepted), the way you see others is incorrect, and the way they see you is also incorrect. We are all wrong, and this is the meaning of Buddha, which all the social systems deny (termed "ignorance").

This path must be made alone not only because of the overwhelming external denials. It is a long way, in which one must make his or her own verification, both inward (directly) and outward (regarding others). Above all, it refers to the need to step out of the framework (alone). It involves concentration and gathering in (against the framework), and it will not form without this "really so" (conviction) – the frameworks, within which acceptances form, are not objective reality, there is no one out there that sees what you see, and freedom is sacrificed for an impossible illusion. With this understanding, concentration is directed toward the evaluation and choice-making process, to see how they form, instead of regarding the frameworks that direct them as self-evident. Everything forms by causes (passive), and so formed the perception of objectivity and activeness, and they have never been true.

The Path

The Foundations of the Mind is, above of, all a logical scientific explanation, which also forms the path to the boat (showing what and where it is), but when it is understood directly that it is really so (our mind is physics) then it is also the boat itself and its destination. Accordingly, Hui Neng's famous saying that "from the first, not a thing is", actually means that all things form in the mind. In the Dhammapada it is stated even clearer – The mind precedes all things, dominates them, and creates them. Hence, the path is passive, without goals, just seeing things as they form in the mind (as they are) instead of distinct external objects (outward, activeness). To attain this insight, one needs to keep balance among the "five powers" and let them do the job. Knowing that all things form in the mind, we can now review these "powers". Needless to say that these terms must overlap, and are not representing separate entities.

Concentration – when the mind stops drifting with the task management process and remains stable, it is in a concentration mode. It is to step aside, for otherwise, the mind keeps wandering with the story, controlled by the framework. It implies realization to some degree that things form inward; otherwise attention will be outward and the story will keep running. Thus, it was said that without wisdom there is no concentration, and vice versa [17]. Similarly, mindfulness is a prerequisite for concentration since it is the way to detect how things really are (inward). This demonstrates the circular and overlapping nature of these terms and the mind in general.

Mindfulness – when the mind is in a concentration mode it can be attentive to the way the "tasks" form from memory. It is a gaze from the side, without involvement in the story. In other words, when the attention is inward it is mindfulness – essentially the same "attention" but toward a different direction (different orientation). The progress in this path depends on the balance between mindfulness and concentration [17]; to be attentive to the evolvement of the "tasks", and yet, not to drift with them.

Right effort – reaching the state of concentration and mindfulness, and staying in it, is made by Right effort. The formal definition is: 1) to stop unhealthy states, meaning, to step aside; 2) to prevent such states, meaning to remain aside. The same pertains to 3) maintaining healthy states, and 4) striving for more. Simply put, while normal effort is to get (bring into the framework) or to be accepted (within a framework), the Right effort is to stay away from these frameworks. It is much easier, more pleasant and sustainable to stay in tranquil water rather than trying to hold forcefully in the middle of a strong current. This is a passive effort, without attribution to the "I".

Wisdom – to know what is true, possible and worthwhile, and gradually also to understand it directly (as the framework loses its power). This knowledge includes confidence in the Buddha's teaching and the teacher.

Confidence – believing in one's capability to discover. Without it, no effort will be made, and the mind will continue in its habitual way.

If you have attained these five powers then you have found the boat – the state of the gaze from the side, which is concentration mode, alone. Maintaining this state mindfully is the sailing, in which this new perspective (away from the framework) is enhanced. The main point here is that this perspective is of the same level as the framework (not under it), hence it forms a contradiction to the prevailing routes and reveals their flaws (contradict the framework). "This shore" in the river crossing tale, is our presence under the framework. The crossing is completed when the framework is done away with.

Gaze from the Side

All insight practice techniques are actually "gaze from the side", and any real examination is always "from the side", for otherwise, things seem self-evident (when immersed in the story). For this reason, simple things might be the most difficult to understand –it seems improbable that these obvious things may contain any flaw or that there may be something more to understand. Consequently, one will normally keep the same old tracks under the familiar framework, and the foundations will never be verified (from the side).

The normal state is of involvement in the story, that is, actions toward real objects (that are related to frameworks), and attributions to the "I" according to the results. "I" is what had been formed in the framework as part of the choice-making structure and the directing of activity (to fulfill choices). Hence, when the gaze is from the side it is not the same "I" which is present and not the same preferences. Not "I am observing from the side", but rather I and the events related to me are being observed – not as a unified object ("I"), but instead as separate sensations and perceptions (that normally are summed up to form the "I"). It is a different angle, without goals and priorities ("from the side"). If it is not the normal "I" that is being accepted, and not part of the hierarchy of task management, then it is not the same compact "I" that is required for such acceptance (and form by it). The same applies to all other objects, that without the involvement in the story do not seem so real, and this is the meaning of letting go of clinging (detailed below). By definition, the "gaze from the side" already means "no-self" since the self evolves in the framework, while now it is "from the side" (outside the framework). Thus, the instructions to observe the body (rupa) siting and walking, or to realize that nama hears and sees (not you) [18], are aimed to form detachment from the story, which is a gaze from the side.

Examples:

The surprise that creates the "overturning" (realization) arrives from the side, i.e., external to the task management hierarchy (hence surprising), and when this structure is already weakened, it crashes as a result.

If "I hear" (I am part of the story) then the heard objects are real and external. If "I am being heard" it is even clearer.

"Turning toward is turning away" (Nansen), because it involves a goal, under the framework – the exact opposite of what is needed for the attainment, which is stepping out of the framework.

Object Dismantling

The objects of reality were formed as summations of features, both regarding their basis and later associations/attributions (of features) to the "entire object". Hence, understanding (insight) involves dismantling the object into its components, which is simply, to begin from the beginning. The Buddha clearly did so in the Satipaṭṭhāna Sutta (regarding the human body), and similarly, the saying that "in the seen there must be just the seen" (Bāhiya Sutta) [5] concerns dismantling of associations; in this case, of the different sense doors. This approach is not restricted to tangible objects since all objects of the mind are subject to the same principles. The objective is to diminish the frameworks in which these objects form and interact, and ultimately, to contradict and weaken the top-level framework. Accordingly, when for you, in the seen there is just the seen, in the sensed there is just the sensed, in the cognized just the cognized... then the framework has been done with, the path realized, and this is the end of suffering (a variation on the original sutta).

The koan "one hand clapping" is another example. The objective here is not the hand, as it might seem, but rather dismantling the familiar association – two hands clapping also cannot make a sound. The sound is created at the "ear base" and is connected to the sight of the hands at a later stage (of the information processing in the brain), due to the correlation of the timing of the sound and the sight. Instead of focusing on an artificial koan, you may just as well pay attention inwardly (i.e. be mindful) while washing the dishes (or any other activity). The sight of the dishes, the movements of the hands, their sensations, the sound of the water, and the clicks of the dishes – they are connected only as associations and only in your mind. What the babies (that we had been) had learned is such associations, that now seem objective and external. Unfortunately, it came together with the distortion of the original balance, and regaining it requires this dismantling. Association is essentially acceptance (where the correlation is best), but it is only a correlation (between neural structures), and not external as it is perceived.

The very act of dismantling an object that just a moment ago had been part of the story (or even part of the framework) is a "gaze from the side". It is not only a pause or a preparation before the real practice but rather the solution itself. While the "stepping aside" relates mostly to the "I" (that otherwise forms as a summation of features which are accepted within the top framework), it permits and creates the review and dismantling of various past associations (that otherwise seem real). This includes associations concerning oneself, which thereby do not remain the same.

Examples:
The face doesn't see but rather is seen, and the seen eyes are present in your mind only, and those that truly exist only

transfer signals to the brain. It enables the creation of seeing in that brain, but it is not you that is seen there, but rather another object is formed, based on a very different memory (and correlating light rays).

Look at your own picture. The eyes in the picture could not see outward, so what is this picture?

At the end of an internal consideration, what was the "I" that was present during it?

Task Management

The whole mental process, with all its aspects, is becoming by means of a "task management" process; hence, recognizing this should be a central element in the practice. This is observation inward, of course. One needs to see and note – all the bodily and mental movements are toward acceptance (not I make them); all the tasks are there to gain acceptance – see it and note. The tasks designed to attain awareness (the practice) are also performed for acceptance, hence they embody inherent contradiction (circular trap), which interferes with progress. Everything is subject to this dynamic, and realizing that is the gaze from the side (mindfulness).

Though it is so, regarding the top level it doesn't need to remain so, and this is the goal of the practice. The dynamics, of course, remain the same, but the structure can change. Thus, while thinking, speaking or acting, take a pause and look – the "I" is being accepted, yet not active. Then, continue, and you will see – the action is still performed well, and it clarifies that not really "I am" doing. Activeness has always been only an attribution.

The tasks run constantly, as we know them, and as the Buddha defined (the "wheel of becoming"), hence, the completion of the tasks (the ultimate goal) is not inherent in the acceptance but rather in stopping the need. That is, realizing with certainty that the framework is unsubstantial, and therefore unsafe. The wheel of becoming, presented by the Buddha, is a gaze from the side regarding a particular "task" – observing how it emerges in the mind, beginning from the formation of the object, the choice in its regard (craving and clinging), and finally the attribution (end of task; death). It is propelled by the framework (ignorance), which is the target of the insight practice (to eliminate it).

Needless to say that a task is made up of subtasks on many levels, yet, the Buddha focusses only on the observable level that is conducive for practice. This is because the aim is to end suffering, not to teach theory. Note that while the wheel relates to the formation of clinging, the awareness of the wheel is, by itself, the end of that same clinging. This is mindfulness – focusing on the becoming process instead of the normal focus on its results, which is attention.

The Bāhiya Sutta relates to this topic as well, although differently, and it is important to see the overlap. It states that when it is clear (regarding oneself) that in the seen there must be just the seen, then it is the end of suffering. That is, "in the seen there is only the seen" is, by itself, the release of clinging, hence, the end of suffering – the end of the framework, of the need to be accepted in it, and the endless tasks (and frustration) that result. The wheel has come to its end, its chains are broken. The release of clinging, in this case, is the dismantling of the object into its basic components (sight, sound, recall, ..), and is not about giving up on attaining it. It is simply not there anymore, at least not in the way we previously conceived it.

It reminds us that all things (reality) are one huge continuous structure (mind), and although the normal emphasis is on activity and choice-making, the "task management" relates to everything. All the meanings form as acceptances in frameworks, and all the frameworks can be viewed as tasks (where the acceptance is the "end-of-task"). It includes all the attributions and associations, which come to their end when "in the seen there is just the seen".

This huge hierarchical structure includes the goal, "I" that is being accepted, a task to be made, subtasks for each movement, associations, the formation of objects, and the formation of basic features (in reverse order). It may be clearer and more accurate if presented as two hierarchies, one for the formation of reality (bottom-up), and one for me and my activities (top-down). The two naturally meet, interact, and form concurrently, affecting one another (overlap). This corresponds to the brain's sensory and association areas (reality; five sense doors) vs. the frontal cortex (framework; sixth sense door), and they meet at the higher association regions, mediated by feelings (limbic areas). As you can see, nothing remains to evoke contradiction or to verify this reality (which forms entirely in your brain); hence we need to cultivate the "gaze from the side" as the ancient wisdom instructs.

49

Returning to Buddhist terms, the nama of sixth sense door is this top-down hierarchy (the framework) while its rupa is a bottom-up summation, distinguished from the other aspects of reality (a separate rupa for each sense door). Since this rupa is distinguished from that of the eyes sense door (for example), it does not include the seen body but rather the way of its activation. This definition fits well the "task management" requirements, though the translation "body" may confuse.

Our expectations and beliefs, too, were formed as associations between events; everything forms as acceptances according to correlations. It means that the objectivity we so firmly believe in and live by, is just a feeling – it is merely that which had been accepted in the past, which now seems to be unequivocal truth. Simply put, the neural (and mental) routes were decided long ago, and this fixation forms the consistency that appears as truth. Missing are the repressed elements that prove the contradictions, and this is why we are here. So far, the main theme has been to show that Buddhist theory and practice coincide with the physics of the mind and with main scientific findings, which we have all accepted as correct; thus, we may attain the conviction that it is Really So (our minds are biased). With such conviction, we can get onboard.

All the "objective" ones will agree that what exists only in the mind is not true, hence, when the arguments are taken from science, all they can do is to close their ears and step away. This book is not for them. It is for the sincere, the truth seekers, who realize that the evaluation and the choice-making process we are so busy with, is too important to be left unverified. Certainly so, if such clarification helps to end suffering.

Example: "I am" the framework to the action, and this is how it is being directed – to be accepted to the "I" (to me), while "I" need to be accepted within the framework above (top-level). At a lower level, the object I approach is a framework to the movement. Without the hierarchy above, and without the next tasks in mind, the focus is on the action itself, and then it is clear – it is performed, but not by you.

Observing the Body

Mindfulness to the body is a major meditation technique, but how does it work? The drive for acceptance within the level above, which characterizes all levels of information processing (in the brain), is even perceived so intuitively at the top level. Yet, it is an abstract unsubstantial structure, which may fit an intangible "I", but not the body. Hence, the awareness of the body prevents this imaginary acceptance and forms a contradiction to the framework, which is the aim of the practice. Thus, it was said that unawareness of the body is the cause of ignorance.

The effect of the body can also be examined from a lower-level perspective. The general drive for acceptance leads to the emphasis on objects, places, and external destinations that permit it. It means unawareness of the body, which only serves the goal. As a rule, the focus is always on the destination, and the body moves accordingly. For example, focus on the glass, and the hand reaches it; focus on the fridge, and the legs carry you there; focus on a word and the fingers write it; focus on an idea, and the words emerge while speaking. Not you find each word in memory, add to it the laws of grammar and activate the vocal cords properly, while creating the intonation that fits the circumstances.

When the focus is on the destination, it is not on the body, hence, although the action is performed well, the awareness of the body is incomplete. The opposite is also true – when the focus is on the body it is not on the destination, and then the task management hierarchy loses its power, the framework diminishes, and the goal of the practice is achieved. In the buddha's terms, a destination in focus is the desire and clinging, while insight is when it ends. The common ground shared by all the events that evoke insight is a sudden interruption of the ongoing internal goal-oriented "task management". It involves awareness of the immediate presence, here and now, including the body, and thereby the previous chatter is realized as unreal. This may happen if the required conditions were established by practice.

While observing the body in action, it is relatively easy to distinguish between nama and rupa. Then, instead of the conventional active "I" that the body and mind are "mine", it is discerned that the mind (nama) activates the body (rupa). The "I" is just a summation of body and mind that is being accepted in the framework, hence it gains the attributions (which make it seem real and active). Note that the "body" here is a representation in the mind, so the expression "mind" represents only part of the big Mind, which includes the whole world. The "I" here can serve also as a demonstration of the overlap and multiple representations that characterize the brain's representation method [21].

Examples:
The hand moves to the bottle while the "I" is only being

accepted in the framework, hence not I activated the hand but nama. However, nama is a framework, it cannot be related to, and transforms into an object (additional level) when such relating is needed. In this manner, the "I" was formed, as an object of attribution. I and nama overlap (hence the error is unclear), the hand and nama overlap ("my hand"), the hand and rupa overlap. All these objects are just summation levels, where the top-level ("I") gains the attribution (I moved my hand). The attributions create active notion although everything is passive. The active note is enabled by the separation (objects), yet all things overlap, and the mental trap (catch) is all-inclusive. This example cuts the "I" into two elements and constitutes another perspective, which is from the side. The normal "I see" is also just a perspective and not the true state of things. There is never someone that sees, but rather attributions that form the perspective of involvement in the story (under framework), and in the lack of anything to compare with, it seems an objective reality.

The question "what carries your lifeless body" (that the senior monk asked his friend) leads to the answer what doesn't carry it ("I") and that "I" and the body are not unified (the distinction between nama and rupa). It clarifies that another person sees you in a very different light (the "I" is contradicted; not accepted) and that his mind is very different than your expectations (the framework is contradicted). Consequently, the friend became enlightened.

The Three Characteristics (no-self)

In the hierarchical structure of frameworks and central projections ("task management"), each level "recognizes" the one below it (which projects to it) from the "outside", as a unified object, disregarding all the prior complexity which enabled it. It means that a subtask is summed up to the one above it as an object. It is so when the focus is on the destination, and then the three characteristics are not seen (the insight is hidden). When attention is on the action itself, without a destination (i.e. mindfulness), the elements are seen more clearly, until eventually the three characteristics are revealed (when the hierarchical structure is broken). It is not a hand, but rather a summation of sight, sensations of the muscles and skin, the memory of related events, capabilities, and ways of operation. If the mindful state is maintained constantly, these elements will also lose their fixation, and the rise and fall will be seen (impermanence; the first characteristic). It is comparable to a movie, in which many consecutive slides appear to be a continuous movement.

In this manner, the "I" loses its validity. Simply, without the acceptance within the framework, the "I" is not an object and its elements are revealed (no-self; the third characteristic). As the "method" for choice making and directing activity (to fit the framework), the "I" is constantly being accepted within the framework (as a unified object), and when the goal-oriented task management structure ceases, so does the "I". Yet, it is a circular structure; when the "I" is there, the action is not seen in itself, but rather as a subtask, under a task, with the "I" on top (under the framework). Circular means that we have a catch here (how to begin), which is quite evident. Hence the directions are general, aiming at gradual progress. Just the "gaze from the side", without being immersed in the story, will do the trick. The meaning of the three characteristics is that there are no objects other than mental formations, and only because of the frameworks' fixation do they seem tangible, real, external and objective. When this is clear, the constant "task management" appears as suffering (the second characteristic).

Example: She said. However, not the mouth said, but rather the voice was formed in your "ear-base", and similarly "she" was formed in another region of your brain, and what was really on her mind is completely undisclosed to you. The "you" in her mind is very different from your self-perception, and her expectations are probably very different than yours. What truly exists are only correlations in many dimensions, and whatever is not correlated with the prevailing routes is repressed.

The Lost Arrow

Summarizing the above, using graphics might be too difficult. However, I attempt to present it here as the direction of an arrow within a table that shows the main aspects of three progress stages (naturally, these aspects overlap). The "normal arrow" is normal goal-oriented task management, propelled by the drive for acceptance in the highest framework, which by that gains the power to direct the lower levels. It concerns attention outward, represented here by the normal arrow (while the circular aspect of driving power is not shown graphically). The "Lost Arrow", which we need to cultivate, is the inward direction, the gaze from the side, and awareness of the formation process (as opposed to believing in the results and focusing on the destination, outward).

In this process, the object always moves toward the framework (destination). Thus, while the focus is on the destination, the legs carry you there; the focus is on the bottle (the destination in this case) and the hand reaches it; the focus is on an idea and the details and words (objects) surface. A novel task requires focusing on a lower level destination (more detailed), yet the same principle applies.

This is the basic dynamics, yet, the destination is not a destination without the hierarchy, in which the higher levels (goals, wishes) direct and dictate the ones below (by the three-level structure). This propels the mental process and forms the desires and clinging, which underlies our normal behavior. In contrast, the Lost Arrow is a change of direction, made by changing the driving force (the framework). With that, the attributions to the "I" cease, and there are no more "rebirths". As a rule, each attribution enhances the framework and constitutes an "end-of-task", after which the "wheel" begins a new cycle. Normally there is no awareness of this natural process; the awareness of this "wheel" is the Lost Arrow.

Level	Normal state	Practice	Insight (result)
Mind sense door	Self , direction is outward, goals and tasks	Direction inward, focus on becoming, not on results	Rise and fall
The framework (that sees)	Following "external" authority. Awareness to the process is repressed. Acceptance is conditional.	Doubt; no authority. Focus on the general structure and processes.	No seer. Nothing is lacking. Acceptance is unconditional. End of suffering (no framework).
The I	I'm OK because I comply with the conditions (of the frameworks that form me).	I am "not OK" – requires review. Spirit of inquiry with faith. Basic confidence and wisdom	No-self, objects do not seem real
Objects and destinations	Seem real and form external demands. I get and I am being accepted	Sensory input is the basis for observation. Everything forms, Nothing is self evident	The destinations do not seem real, external or valid.
5 sense doors	Process is repressed, nama-rupa conjoined	Nama-rupa distinguished	Rise and fall

For a summary, it should be appropriate to relate to the original, well-directing, top-down overview called clear comprehension (Sampajañña): 1) Purpose, 2) suitability, 3) domain, and 4) reality. The top-level is the purpose of the action; it should be in the best interest of oneself and others, particularly with regard to enhancing the understanding. Then the action's suitability should be considered, as already mentioned above. Having done that, the actions are well directed, and so should be the involved attention (domain) – all experiences are made a topic of mindful awareness. The fourth aspect reminds you of the proper goal, which is seeing reality as it truly is – all actions and perceptions stem from causes, are impersonal, and in constant change. This last aspect is the main topic here while trying to provide a scientific explanation (based on physics and neurobiology) to our mental content. The main question in this regard is: With such wise directions (of the Buddha), why still most people fail to comprehend? My answer is that the clarification of the sixth sense door and its biases (framework) were still lacking.

To demonstrate this point, note the overlap: "Clear purpose" is to find out by yourself; the suitability means that others cannot help you (hence, you must find by yourself); the domain is a research that you can do only by yourself; clear comprehension of reality is what you find out when you begin from the beginning and research the foundation by yourself. The list of distinguished items (due to the attribution's bias) become aspects of the same thing.

Clear comprehension is achieved by practice [19], that is, for a true comprehension of purpose and suitability it is required to grasp correctly the nature of the objects – overcome the essence of ignorance. Then, also clear comprehension of the domain and reality are possible. For this aim, most of us need a proper explanation, which provides a starting point and facilitates the observation.

CONCLUSIONS

Already Arrived

Meditation is a mental structure, and if it is aimed at achieving a specific goal then it is the normal structure (not meditation). This involves an inherent catch (circular trap) – if you believe that you have arrived, you will not observe, and without this attitude that nothing is lacking, you will not progress (the necessary structure will not be present). It is manifested, for example, in the instruction to reach the destination with every step of the way [3], or the warning that "when you turn toward, it turns away" (Nansen), etc.

Progress here is against the goals, and it means alone, outside the framework, hence it doesn't feel fitting or desirable. Release of clinging (or attachment) sounds like giving up or losing, and it doesn't "fit for me". The gaze from the side, on the other hand, is a release from the framework, emancipation, and this is actually the release of clinging. It is the solution because it acts on the framework (diminishes it) instead of the feeling of isolation if alone while the framework is dominant.

The way is made by adhering to the truth in the most accurate and complete manner possible, because otherwise, the mind has goals and an "I" (under the framework). The ultimate truth

61

is attained when all things fit together, as opposed to the normal approach in which everything must fit the framework. Again, "already arrived" is an essential attitude, when the goals are less dominant. In any case, it remains circular, never objective, but the balance and quality are regained. For example, when explaining to others, the framework prevails, and the spoken sentences are understood according to the context (not understood). However, when the readiness is there, a surprise sound may evoke the awakening, because it clarifies that it is "really so", and not only a theory, to be told to (or by) others. The flaw was not in not achieving something, but rather in the need for it (the framework).

Example: Freud's repetitive compulsion is a behavior that cannot fulfill the expected acceptance. It is created by the framework and repeats because one cannot question a task when dealing with its subtasks, and the supertask (framework) can never be satisfied (hence, the repetitions). Recognizing that, is being cornered. Stepping aside is the surprising solution – in contradiction to the main goal (framework).

To summarize the above, "already arrived" is enlightenment – you are no longer under frameworks, no tasks that must be fulfilled, and no self that should fulfil them. So, to clarify this "arrived" let's review the self, what it is comprised of, the way it forms, and its functionality. As a rule, we look for a simple regularity even when the results appear complex and diverse.

The simplest definition would be that the "I" is a summation of the top levels of all aspects of task management: How my body should move at the present activity, and what "fits for me" regarding more distant and abstract objectives. This defines also what doesn't fit me and I should avoid, even in the face of social demands. My belongings are also part of the definition "what fits for me". The belonging defines rights to use and capabilities, and these are part of my features. My past activities, too, are part of my features since they define my capabilities and what "fits for me" in the future. Accordingly, the "I" is the definition of both the objectives and the ways of realization (e.g. how to move the hand, can I pass through this door, the step is not too high? etc.).

This understanding (of the "I" as summation) means that the common conception "my will" is an inversion between the cause and the effect as part of a fundamental misconception of ourselves and reality. The will is not a separate entity, which is in my control, but rather it is an integral part of the summation called "I", and it connects between me and the desired object (shared feature). All the objects are summations of features, and the "I" is distinguished only by the kind of features that it sums up – those that are relevant for directing the activity (task management).

Each moment different aspects of the "I" are more relevant, and so the "I" (the active summation) constantly changes according to the circumstances. In the lack of tasks there is no "I", and while tasks are fulfilled, the "I" gains the attributions, which essentially are additions or adjustments of features. An attribution is part of the task management structure. It takes place during the activity, and not only at its end. Probably, this occurs at the ad-hoc level and retained in memory in the entorhinal cortex (EC). A detailed discussion in this regard is presented in The Foundations of the Mind. In short, the perception of "time and space" and understanding of new states, which involve new combinations of inputs, require a dedicated memory region. All evidence points at the hippocampus as the engine for the formation of the required combinations, and at the adjacent EC as the region in which these new connections are retained. The "I", which constantly copes and interacts with reality ("time and space"), must arise in that same region. This is a general definition, which relates to all objects, and not only regarding attributions to the "I". Accordingly, the attributions' data is the main cause for our need to sleep (free this dedicated region), and this, too, is elaborated in The Foundations of the Mind.

When you view things in this manner (how they form) nothing seems substantial. This understanding is essential for progress, yet, normally hidden from laymen. The sixteen stages of progress (yana or nana) may give you an idea about that. For example, the Vipassana Dhura Meditation Society writes explicitly: "Generally speaking, we do not recommend that beginners read this article". This reflects a common view among meditation teachers that such information should not be made available to the general public. However, the seemingly naive and simple meditation practice is about observing things of the mind and body as they form, which eradicates the belief in any substantial thing, and creates a mega-revolution. Here it is only an explanation (of how things form in the mind) but it points in the same direction and may help to attain the required conviction.

Course of Life

There is an inherent insincerity in a mind that arises by the drive to obtain (things) and be accepted (within frameworks). It prevents the "beginning from the beginning" and doesn't allow the verification of the framework. Thus, it also embodies a fundamental misunderstanding – we really believe that THIS is an objective reality, and the framework is this belief (yet, in each mind there is a different THIS). The knowledge we worship is nothing but a structure of frameworks (a representation) that might be biased. Therefore, it may not mean understanding, which is the balance and correlation among all the relevant frameworks (without repressions).

Intelligent and stupid are a common combination because stupidity is in the fixation of the frameworks while intelligence is in the use of them (and knowledge is their existence). Stupidity is manifested in a wrong view about the self and the world, by which all the interpretations, evaluations and choices are dictated. One needs Concentration and Mindfulness in order to see such basic misunderstandings, and this means a change in the frameworks. The bigger the change, the more Wisdom, Confidence and Right effort that it will require. Without it, intelligence is activated by the prevailing frameworks, which had been established long ago. The motivational catch (trap) was already stated earlier, and to make things worse, there is an inverse ratio between the need and the capabilities.

Life does not wait for anyone. Each step dictates the next one and was dictated by the previous. I sat in a certain place, and this dictates what I see and what events may occur – it dictates reality. Why did I come there in the first place? Because there was a previous reality, and choices were made accordingly. The small steps dictate also the bigger ones, although the frameworks maintain a general direction (for good and for bad).

A person progresses in areas in which he/she experienced acceptance, and thereafter accepts others in those same areas. It is a fractal structure in which each object is also a framework to others, dynamically changing each moment. The "I" was defined (learned) during the course of life, and every moment it forms in the mind based on memory (knowledge and frameworks) and affects the choices and courses that are pursued. There is nothing substantial or real in this regard, though this is difficult to accept, and normally must not be said. However, consistency proves it, and there is no other measure for truth but a complete consistency (correlation).

Reality is forming as complex bottom-up summations within the sensory and association areas of the brain. The evaluations and choices form as a top-down hierarchy, based on this reality. In the lack of anything else to compare with, this reality seems external and real, and the evaluations seem to be objective. Nevertheless, the "direction of sight" proves it wrong, no matter how many people believe otherwise. Science has shown it clearly, yet the problem lies in the bias of attribution solely to the object (that is present in all formulations). This hinders the understanding that it is Really So – regarding the concrete objects of one's own reality, and not just a nice or useful expression.

Reality is similar to a dream; although its objects do not change so wildly, they could have been perceived in other, very different ways. This is not allowed even to be thought, and certainly not to say. The others and their opinions are authority, and you need them, hence, you must comply, flow with the current, even when it dictates a severe error. Nonetheless, each mind is made of objects that remind one another, and you have no way to grasp the objects of another mind or to predict their outcome. You can understand only alone, and actually, this is the understanding itself (away from the framework).

Stepping Aside

To summarize the path, progress is formed due to the state of mindful attention to the way the stories evolve, instead of drifting with them (as if they were an external objective reality) or alternatively, fixating on one object only (when concentration is too strong). This simply means a balance between Concentration and Mindfulness (please see the prior definitions). In order to achieve that, all you need is Right effort – to step aside and remain in this state. However, for the right effort to emerge (instead of normal effort) you need Wisdom and Confidence in your capability. Both are enhanced by Concentration and Mindfulness. Obviously, this is circular, yet after obtaining the basic knowledge/wisdom (as in this book), I would put the emphasis on the right effort, as the engine of progress. Here you need to make it concrete, regarding your own specific stories and frameworks (to be stepped outside of). If you take it seriously, regarding all actions and thoughts, and all the time, you will surely see results. But this is a big "if", hence, you may need a retreat, with friends (sangha) and a teacher, just in order to gain this seriousness.

The four noble truths are: 1) suffering, 2) the cause of suffering, 3) the end of suffering, and 4) the path leading toward the end of suffering. Suffering is the framework (which is also ignorance); the cause of suffering is the clinging and craving that stem from the framework and maintain it (circular); and the path is described in many books. It is the third truth that seems strange, you may say redundant or obvious, yet, it is there for a reason (and not after the path as a final result). It refers to basic wisdom and confidence, without which the right effort will not persist and the path (fourth truth) will not be entered. It all comes down to conviction [in the correct direction], which is also the aim of this book – based on scientific facts and pure logic (consistency) to build a synthesis, which provides the confirmation – it is Really So (attaining the basic wisdom). The third noble truth is the Buddha's "Really So" – the actual stepping aside (internally) – the key to the boat (path). It is called "end of suffering" because the framework is the suffering while the third noble truth is the stepping out of it. This is normally overlooked or simply not stressed enough, yet, without a key, you cannot enter.

Final Notes

As a final note, just keep in mind the five powers, the role of Right effort in that, and the conditions for its emergence, and you are already set to be free. To begin with (get on board), the focus should be on the framework, since diminishing it is the objective of the practice. I am always under a framework and within the world, the objects of which are [apparently] external to me. Every moment another object is selected for the primary interaction with me (thus it becomes part of the framework), and the mental activity is directed accordingly. This constitutes the "structure of interaction", which defines what I should or must not do, how, and the expected results – the essence of the framework (craving and clinging, in the Buddha's terms). It is also the structure of the attribution to the "I". This structure is always there while the activity is going on, and upon completion, only an "end mark" is added.

When being observed (from the side), this construct doesn't seem so real. It is a mental structure, not at all objective, and without its imagined acceptances and attributions to the "I", I too, do not seem real. It wasn't me that was speaking, walking or deciding. It has always been the structure of "task management" that was making all the choices ("what fits for me"), acceptances, the illusion of activeness (the attribution "I choose") and the automatic rejection of everything that doesn't fit itself – thus maintaining itself and its fundamental error. All these come to an end when "in the seen there is just the seen", that is, when the "structure of interaction" (the framework and its attributions) does not arise in conjunction with the seen or heard, etc. (detachment). And then "there is no you there" [5]. This alone is the end of suffering.

It is relatively easy to see that the top-down hierarchy of evaluation and choice-making (structure of interaction) is completely internal, and the reality it relates to must be in the same place, though our intuition is very different. Accordingly, concentration and attention should be directed toward the structure of the current interaction (whether imagined or real), in order to see its structural flaws directly, as opposed to focusing on the objects and taking the structure for granted (normal attention). This is the difference between wisdom and intelligence and between inward and outward. It is "from the side" and it weakens the framework.

Without the control of the framework, the chatter stops, and this is "concentration". Attention in this state is "mindfulness" (inward). It is called the "knower" or "watcher" state, which is the "right concentration" [19]. The watcher is stable in the background, distinct and detached from the observed phenomena, which constantly change in the foreground. This corresponds with the instruction to observe the whole posture and not focus on any specific organ [18]. It is like watching someone else from the distance. Not "my" body, not "my" thoughts.

Normally, the mind switches between the "thinker" and the "watcher" states. It cannot be both at the same time [19] since the "thinker" is under frameworks, while the "watcher" is the gaze from the side. If the "knower" (or "watcher") state is achieved, you don't need to watch your breath (concentration practice); you are already onboard (of the "boat"). Just keep on in this state and the "boat" will take you to the other peaceful side of life.

The "letting go" of the structure of attribution to the "I" (=structure of interaction) is letting go of the conventional reality. That is, when "in the seen there is just the seen", images are seen as images, neither approved nor condemned (without the conventional meanings and relations). The conventional reality forms within the framework and collapses together with the framework. Neuroscience has clearly shown that in the processing of sight there are two main distinguished pathways: 1) the "what", and 2) the "where", and they are integrated to represent the perceived objects. However, this is a flawed interpretation. It is not just "processing of visual information" but rather it is God's work – the creation of the world (ultimate truth instead of conventional reality). The "where" is forming in the brain and exists in the brain as a feature of the "what", which forms in the brain concurrently. In every instant, the whole world is created in your own brain. Real magic. The small things are not at all small as they seem, and the big ones may not be as great and important as they appear, so we may relax and enjoy the present.

The buddha's roadmap, termed Clear Comprehension, is finally clear. When the objects seem real and objective, so do the frameworks in which they appear, and this is the wrong comprehension that causes the wrong purpose. In contrast, one who sees that the objects form in the mind, is not looking for confirmation or acceptance, not to control or possess them, nor belong to them (1-clear purpose). This is about the structure of interaction that has changed – instead of being involved in the story you are now watching from the side. Then, it is also clear that what the others perceive is a result of frameworks behind their eyes, and it is very different from your perception (2-suitability). This pushes the mind further toward observation mode, which is from the side (3-domain). Only if this state persists can you eventually comprehend the truth (4-reality).

To attain clear comprehension is to penetrate the third noble truth (stepping onboard). Otherwise, while the conventional structure of interaction prevails, it will not help you to observe the body [see 19], and the structure [of interaction] itself should be the subject of observation, in order to find the key to enter (which is the stepping aside of this structure, as suggested by clear comprehension of purpose).

The four aspects above are normally presented as a list of tools for your practice, but then you miss their essence. There are no tools except for the observation from the side, there is no one that can use tools, and there is no better destination other than escaping the tyranny of the frameworks and their "task management". Note the catch (circular trap) – using tools is being under the framework, hence the "tools" in this case, prevent the work (of ending the framework) from being done. However, you need a good roadmap in this illusory and deceptive world, and need specific directions in order to stay on track. The famous eightfold path is normally regarded as a "roadmap", yet, it relates to a long period, while clear comprehension is about navigating your mind in the present moment, without which, the years will pass and the eightfold path will not be accomplished.

I try to make these abstract terms more tangible and concrete to have them usable; hence, I repeat them from various perspectives. Observing the structure is the key and is the gaze from the side. While the structure is observed, the attributions do not seem real, the activeness appears as unreal (as attribution), and reality as a representation, which normally seems external as result of the observed structure. This clarifies the term "ignorance" (related to the believers in reality) and constitutes the departure to new life.

It reminds me of the focal question, why things were not explained? The belief in activeness (as opposed to causality) blocks any explanation, and the belief in external reality (objectivity) is also the belief in the attributions (as real), including those of activeness. A circular structure that maintains itself, apparently a trap with no way out. Here comes the Buddha, the bright sign above the exit door, showing you the escapeway. Instead of the normal attention (to "external" entities), in which the attributions seem real, you are advised to Concentrate (to stop this wild run) and direct the attention toward the driving force (to be Mindful) – a change of direction (that reveals the causality and diminishes the framework).

But how to do? To put your effort in context and make it effective you need to follow the roadmap (remembering that the law of causality always applies). Hence, I repeat once more:

Clear comprehension of purpose is attained by Right effort and sincere mind, realizing that it is Really So. The new direction is your purpose now because the old one has failed. Without this realization (being cornered) you will not break through. Simply, there are only two main directions, and if the normal one is correct, the other one is impossible. It is the stepping aside. Clear comprehension of suitability complements the stepping aside. While you attribute objectivity and authority to others, the normal direction is dictated, hence you need Right effort to stop it. With that in place (your goal is to attain direct understanding and the others do not seem as great help for that), your Concentration and Mindfulness practice is effective (clear comprehension of the domain). Clear comprehension of reality is the result. It completes the Wisdom by which you could begin this path. Reality has always been behind your eyes, and only partially correlating with that which exists in front of them (a representation).

The essence of this path is passive observation because activeness is the main illusion. It seems real while the truth is that the existence of a goal (framework) is the feeling of activeness, the origin of the compact self, and the attributions of the actions to it, as part of an automatic directing of the activity (according to that goal). The normal belief that we are active is simply a confusion between the cause and the result. It stems from the nature of the objects, which are summations of features (results) but seem distinct and real (causes). This applies also to words, of course, which makes this explanation difficult to formulate (e.g. the objects "activeness", "I", "will", "goal", "acceptance", etc.). For example, the acceptance of the self in the framework (goal) is the activeness (feeling); the acceptance is the attribution. Nothing is distinct here, yet, activeness means that I am the cause, hence separated, which is a wrong understanding. There are many words but they are all overlapping summations; not substantial entities. All aspects of reality arise from memory similarly (whether the recall is based on sensory input or not), and this renders our normal estimations and choice-making far less accurate and worthy than it seems. Most notable is the "I", which is nature's "patent" for choice-making. When it is understood as a summation of features there is no activeness, no acceptance, and no framework, and then the mind opens up – a unique and indescribable experience (enlightenment).

This is the realization of the fractal laws – the "I" is realized as similar to all other objects (not active or center), and the will is no longer a great command. Activeness has always been only an attribution, and all objects have always been representations that summed up various features to seemingly compact entities, and all this has never been external to your mind, generated by the natural fractal regularity (not you). Consequently, there has never been anything different to compare with, and the biases seemed objective truth. Most difficult is the idea that you are being seen (main structure; the framework), which renders everything as external and real, and underlies primary goals (of acceptance). Hence, stepping aside, concentration, and mindfulness are certainly required to form this shift and realize the truth.

With that said, we can return now to the sixth sense door, which is the key to insight while the framework prevails (thinker mode). You may view the explanations here as belonging to the sixth door, yet they remain under the framework. Missing is the stepping aside (away from the framework), and this, too, can be more reliably achieved via the sixth door, thereafter, also the other five doors are more revealing. The suggested practice is derived directly from the explanations above. For sure, it remains passive – just noting regarding the nature of the phenomena that have arisen in the mind. The difference is only about the aspects you focus on. Accordingly, whenever you have a moment of awareness (a break in the ongoing task management process), note the following:

1) It seems that I am being seen (framework); 2) There is a feeling of acceptance or effort to be accepted; 3) It feels like being active and center; 4) It seems that I am choosing well (what fits for me); and 5) I feel objective. However, 6) all these are results of causes (in the brain), are aspects of the framework, and present only in my mind (specify each one).

This practice is about observing the structure [of interaction], and applying it consistently will gradually and surely diminish the framework. In a way, it is similar to the Buddha's focus on the wheel of becoming, but it relates to a later stage – after you are already caught up in clinging and before entering the next cycle. Hence, it constitutes the "path to the boat", which takes you to the starting point, in which awareness of the earlier stages of becoming is possible, and the practice of the five sense doors is effective.

We can now conclude that activeness is the gist of ignorance (and not only a bias attributed to a certain mental object) – the unawareness regarding the essence of one's own being. A mental object (self), which is created by attributions, seems substantial and self-evident, and the attributed features seem real, in the lack of awareness of the attributing process. Unfortunately, this state concerns also those whose role is to reveal the truth (science). They will keep promoting their "value categories" even though their own findings completely disprove these ideas of objectivity and activeness. Being part of objective respectful science is a major element of their value; it defines how they should behave and what they should or shouldn't say. This is a religion and it is unquestionable. Thus, they are misdirecting, and we are all trapped in a deep illusion.

If you still have some doubts about whether this "stepping aside" is appropriate, just take a look at the "direction of sight" drawing. The facts are very clear, and although our normal intuition is stronger, the facts are much better verified and they belie the conventional intuition. Four powers of insight were mentioned above, showing how everything connects. The fifth power is Confidence (in your capability). It is the starting point. Having that, you can begin, proceed, and complete this path. Hopefully, these pages show you how simple and logical all this is, enhancing your confidence to observe.

The "ultimate truth" means that nothing can remain (everything constantly forms anew). That is, impermanence, suffering, and lack of self, characterize all things (the formal definition). Unfortunately, this may mean that we are wasting in vain most of the precious time that was given to us. The Buddha is a wake-up call.

APPENDIX - Illustrations

The water is running out of the tap. We can distinguish between a few different features: Location, form, color, and motion (and everything you know about water). These features are summed up to form the picture as you see it. For example, there is a neuropsychological disorder in which the running water seems like ice (static). In this case, the motion feature is not summed up as normal. This testifies to the normal state – what is it that we really see? The answer is simple – seeing is evoked by input to the eyes, as we all know, but what is actually being seen must arise behind the eyes, and not where it seems to be located.

The flickering lights on the screen seem like conscious active figures because they are connected to reality. However, you may reverse the order and conclude that reality has formed in the mind similarly, only at an earlier stage; now it is an unquestionable authority, mainly due to the great correlations concerning time and space, overshadowing the poor correlations in other dimensions.

The "structure of interaction" refers primarily to the principles of the "task management" (of acceptances at many levels of frameworks), yet, it can and should be detailed regarding various aspects. At the top level, it is easy to discern that the "I" is being accepted within the framework while nama activates the body (not I do). "Nama" refers to the framework, including its lower levels, which are organized as a "task management" structure as stated above. A significant feature of the top-level framework is "your" will. Within an interaction, the will of the other (as represented in your mind) is also part of the framework. Thus, the difference between "get" (bring into the framework) and "be accepted" (within another framework) is simply a question of whose will is more dominant, yours ("get") or of the other ("be accepted"), and both may prevail concurrently. The same principles apply to all levels (fractal), including the way the objects in space (and the more abstract objects) get organized (i.e. identification and interpretation). Each object is made of features, which may serve also as a framework for the other objects in the interaction (shared feature). In this way the definition what is included in what (or what is a framework to what) is created. This is also the way by which objects are recalled (by their shared feature) [21]. Under this light, it should be interesting to reevaluate the findings of Gestalt psychology and provide them a simple and consistent explanation (to the way meaning is concluded).

When thinking, the objects around you are not seen (or only vaguely), and when the thoughts stop, the immediate reality is constructed in your mind. It is created with all its meanings, including locations and the [illusory] idea that it is an external objective reality. Yet, most of the meanings arise from memory (by neurons), in a similar way that thoughts have arisen.

82

When the observation is from the side the attributions are not to me because "I" am on the side, detached from the events. If this state persists, you discover that really it wasn't you that was doing, and really your activeness was just an attribution, based on a fixation, which maintains itself.

Please note that the terms "structure of attribution to the I", "structure of interaction", "within or under frameworks", "involvement in the story", "acceptance" – all overlap, only the emphasis vary. For example, the term "structure of interaction" clearly relates to all the relevant objects throughout their interaction, while the term "structure of attribution to the I" focuses only on one object ("I") and the final result of that interaction. The structure of attribution (to any object) is made by frameworks, while the need for acceptance within them is their essence. This can serve also as a demonstration of the bias of attribution solely to the object while other, overlapping objects (the terms above) seem so different. This dynamics for acceptance within frameworks creates also the objects themselves, and not only the interaction between them [21]. Thus, the belief in substantial objects is the lack of awareness of their formation process, of their structure, and of the structure of their interaction – the essence of ignorance (mentioned earlier). This we refuse to admit or verify – because we, our beliefs, perceptions, choices, and actions are the [biased] results, which will be falsified by true observation (from the beginning).

The seeing and understanding emerge in the mind as a manifestation of acceptances within many levels of frameworks. In this manner, the wills (highest framework) affect the features of the objects and the way the objects are understood. So it is with regard to inanimate objects, words, people, organizations, ideas, political views, beliefs – all reality forms according to established frameworks. This has an existential aspect – the resulting understanding is according to "what fits for me" and not just a mere truth. It involves significant insincerity, as we all know, yet, it also embodies misunderstanding. The deceptions and errors are mixed. Even the existential benefit is questionable (or limited to a certain range) since the flawed understanding results in pointless struggles, failures, frustrations, suffering, and occasionally significant damage (and even real risk). When awareness of this fundamental flaw begins to emerge, then, for example, instead of "I didn't understand her" one would say "I didn't understand what she is" (a more basic level), yet, it also means "I didn't understand what I am". I and she are overlapping mental objects. To begin with, what "she sees" is a projection in my mind, and then, her seeing is a framework to me – it defines musts, possibilities, acceptable responses, etc. This overlap means that if I am not real, then so is she, and vice versa.

Similarly, "In reference to the seen, there is only the seen" [5] can be also understood as "what is seen is seeing" – referring to the basic level of what is an object (when the framework is deactivated). We can now add to the Bāhiya Sutra "when there is no you there, no one else is there", which is the meaning of the famous "no-self" (no objects). The great authority (she, he, they, ...), due to which you need so much, is just a framework in your own mind.

The object you interact with is part of the framework. Without the framework (structure of interaction) that object is different, and so are you (the interaction is a shared feature).

Drying the hands after washing them, and before preparing a cup of tea: The tea here is an additional task, hence the "hands drying" is under a framework, and lacks the spirit of Zen [see 2]. The cup of tea disrupts the quality (of Zen), and without it, only the action is present in the mind, here and now.

The young and lovely face did not match the words, the vocal play that followed did not fit, the volume did not match the tiny figure in the center of the huge concert hall, and that facilitates the realization that the voice doesn't arrive from the mouth, even when it is shown in a close-up on the screen. Its details, like all other details, obviously arise from the observer's memory.

The eyes are seen, and their meaning as seeing arises from memory; the mouth only opens and closes, and the idea that it makes voices arises from memory, hence it is not understood as it truly is; the people in your life seemed accepting because so it is represented in memory, but it was only conventionally so.

The "sailing" (on the path) is not distancing because there isn't from what. This understanding is the distancing and the sailing, and it is all so near.

What is untrue here? There isn't the freedom not to do a thing, not to go anywhere, and not to be anyone.

This section is not complete. It can be made stronger (so you will perceive directly that it is Really So), mainly with regard to "trivial" notions and behaviors. However, as stated earlier, you need to make your own examples and form your own path. Do not fall into the trap of the bias of attribution solely to the object. That is, each example can have many variations, and you should find those that reveal the biases of your own mind, particularly regarding the self and the framework (sixth sense door). The point here is not just finding or recognizing, but rather the direct feeling that "it is really so" (biased), hence evoking change. Anything else can only be a pointer, and I believe and hope that I have provided a good one here.

APPENDIX - Reorientation

The attached letter was sent to about 20 professors in various relevant fields. It explains the rationale, demonstrates and proves our major misconception, and directs to the solution. I believe it is well written, and you would expect that professors will relate seriously to something of this kind, yet, it did not happen. They refused to respond even to my second and more friendly approach. Not even one word, they scared away.

I admit that my writing style is not always appealing, and normally you wouldn't like someone else telling you about yourself, and certainly not in such a decisive manner. However, if it is correct, and we are truth-seekers, then we should wish to review and admit our flaws, for our own sake and for the sake of others who listen to us (especially if we are scientists). It is normal for us to have our minds flawed, and it should be normal to be in a reevaluation state.

Sadly, the philosophy of science is "live and let live". Science is not the truth-seeking enterprise we are taught to believe. It is a business, there are people there, they have needs, and they need to build and keep their positions. It is about their survival, and the system demands a constant supply of new ideas, even if everything had been said already – there must not be a final solution, in contrast to the ultimate goal in life (which is the theme here). Pirsig called it the "church of reason", which its founding charter is "to find and invent an endless proliferation of forms …". It is not clear and sounds exaggerated in the light of the unshakeable authority of science, until you scrutinize the details. Then you understand that it is really so, while the truth is in the simplicity and unity – when the overlap is clear, and all things are well connected. Hence, unfortunately, with the goal of multiplicity in mind, the truth is far away.

The said letter:

Hello, I wish to interest you in my work and hope that you will find the following relevant and important. It shows a fundamental error of science (and all of us), due to which the mind was not explained, and that realizing it may change our outlook and affect many things (for the better).

While it is very simple and easy to verify, the conclusions are extremely difficult to digest (the reason why science failed in this regard). Yet, it requires only a few seconds to get the point. No mathematics, only the main direction. Please see the attached illustration:

The Direction of Sight

Actual state: **Conventional understanding:**

Physical direction   Intuition

@ - The seen objects

As you can see, although seeing is evoked by input to the eyes, as we all know, what is actually being seen must arise behind the eyes, and not where it seems to be located. We are all wrong in this regard, and this is the reason why science failed to explain the mind – they were trying to solve the wrong question – how we see the world, while in fact, the world is already the result of the seeing.

The facts are undeniable – the information goes from the outside to the eyes and inward; never outward as it appears, meaning that the entire world must be your own mind, and each mind is a different, yet correlating, world. Correlation is the key factor here, in all dimensions. It is also the essence of the brain's activity and the way all meanings and actions are formed.

This corresponds with the Buddha's teaching, Hume's and Kant's theories, Einstein's sayings, and many others.

89

Our intuition, on the other hand, corresponds with the psychological findings that we extremely tend to conform and follow authority. Consequently, the obvious facts are not understood due to our strong and contradicting (and flawed) intuition. This intuition was established in the newborns' minds, and now as grownups, we fail to review it. Thus, the emphasis here is "Really So" – admit the known facts and adjust your beliefs.

Throughout history, science progresses via paradigm shifts, termed scientific revolutions (T. Kuhn). Here we are facing the biggest one since it does not pertain to a specific field and not only to science. The most fundamental paradigm is also the simplest one, and the most difficult to digest. Yet, it may be the time now to verify the foundations.

A very concise presentation is found at:
http://www.amazon.com/dp/B08245VPTW
It discusses the actual process in the brain, defines the biases, and solves major issues concerning perception, memory, language, psychology, and the Buddhist path. All are explained by only one simple fractal construct, based on neuroscience and the most fundamental drive toward balance. The main thing here is consistency, and this is the only way to verify anything. The details, analysis' guidelines, philosophy (Quality) and implications are elaborated in the main volumes. However, the main direction presented here is the main point, already proved above by trivial facts.

Note the trap: while we constantly try to solve the issues of our world (whether of real-life or as explanations), the problem and its solution may lie in the essence of our mental content, which, by definition, must embody biases (yet, seems objective).

Please keep in mind that this is copyrighted material, and a link to the books is required if you use it.

Thank you! I will be glad to receive your comments.

U. Notmi   @UNotmi   reallyso.org

Final note

The professors' refusal to relate to this letter should not be surprising. After all, the data is simple and known for many years, and its meaning is consistently avoided. Although simple, it is difficult to understand because it is about overcoming the framework.

As detailed in the "Foundations of the Mind", the driving force is always "to get" or "be accepted" within a huge fractal structure (the brain and mind), and for that aim, scientists need to conceal their true drive (get/acceptance) and pretend that the truth is their main objective. This conflict is most severe when they need to explain the mind, which is also about themselves and their drives, exposing the falsehood of the objective message. In this regard, the authority of science is the problem rather than the way for a solution, as pretended. Yet, the system protects itself, so Pirsig was hospitalized and was given electric shocks, to put him back in order.

Now, reflecting on the hundreds and thousands of profile pictures on the universities' websites – what was it that I saw there? There were only colors on the screen, while I saw authority, faces that look at the world objectively, and a joined effort for revealing the truth of nature and life. It was my own illusion, similar to that of all the people I knew. It wasn't even colors on the screen, but rather the screen, the room and myself were all mental representations – the magic of creation.

This magic corresponds to Pirsig's Quality. It is explained in the "Foundations of the Mind", which also clarifies what is enlightenment, based on the same principles. These show you why he made a lifetime effort against the system (the framework). "He couldn't follow any known method or procedure to uncover its cause because it was these methods and procedures that were all screwed up in the first place. So he drifted. That was all he could do." - Robert Pirsig, *Zen and the Art of Motorcycle Maintenance*, p. 115

Pirsig saw the main problem in the dissociation of ourselves from nature, which, in simple words, relates to the belief that our mental content does not stem from cause (objective), hence ignoring the biases. This is also a bias, which stems from causes, of course. The professors, while students, chose their field because there seemed to be acceptance, they progressed because their writings were accepted, and it happened so because they have accepted the prevailing lines of thoughts. All with the idea to get for themselves. Nothing is objective in this path, nor in the life that preceded it, and the same regarding anything they may do or say now.

May we all be free of authority, and enjoy the Quality in this short life.

APPENDIX – Our Religion

Synopsis – Humankind tends to follow religions, and this must affect the progress of science, or lack of, when the true answer conflicts with the prevailing religion. The term "religion" is used here in a general sense and has nothing to do with a divine entity or common classifications. It refers to a must-follow belief, which directs our perception and choices, without being seriously verified. The consequences of this tendency are often severe, hence, although the focus here is on the main subject that science did not solve (our minds), it has far-reaching implications also in other fields.

Religion is a conception about self and the world. Another religion is another conception, which means that they are all misconceptions. Again, it doesn't refer only to the common classifications; it focuses on the prevailing mental structure that [almost] all of us share.

Frameworks

We live in the world as we know it, each one of us holding to different beliefs. While the normal attitude is that we are objective, the fact is that beliefs do not tend to adjust to evidence. Much trouble in our world is the result of disputes that, allegedly, could have been solved by a simple discussion. Most notable are the religions, which normally are learned while being infants; yet, they remain dominant in the mind for life, evidently, without a true verification. Since religions are quite prevalent, this indicates about the nature of our minds, as opposed to a distinct phenomenon. The regularity that underlies our minds is the main theme here. It is the main subject that science did not solve. Why so? This is the focal question. There are a few reasons that can be pointed out and will be presented here briefly, beginning from our tendency to follow religions.

There must be something inherent to the brain's structure that creates religions and blocks the verification of their basis. I call it frameworks. This is not about a certain distinguished component, but rather, all mental content arises within complex hierarchies of frameworks. All frameworks, at all levels, function similarly, which makes it a fractal structure. Simply put, each framework accepts the elements that fit it and rejects others. The term "religion" refers to the top-level frameworks, but their function is essentially similar to that of the lower-level ones – a true fractal model. Such regularity is a must for biological systems, and certainly regarding the brain, which all of it is made of similar units (neurons).

Since "our religion" is the top-level framework, we all share it similarly and force it on one another. This framework is basic, before any specific belief. We call it proudly "objectivity" and constantly ignore and repress the abundant evidence that disproves it. This is a typical function of any religion. Objectivity is so mighty that we dare not see it as a belief. We must accept it for granted (that our reality is objective), and this is a grand error and a foundation of many specific errors and deceptions. Please note, this is not about someone that is repressing or ignoring but rather it is an automatic result of the neural network. The belief in activeness, that someone is doing (repressing, ignoring, thinking, etc.), is also a result of the neural activity. Both beliefs (objectivity and activeness) arise together and are aspects of the same neural structure.

In this article, I leave aside endless related topics and limit myself to the focal question – why science did not provide us a comprehensive solution to the brain-mind puzzle a long time ago. There are various overlapping reasons, where the top one is the top framework – objectivity. If you are objective then things are clear to you and there is nothing you should check about their basis. Thus, the true verification is blocked. Naturally, many people will claim that they do check, so please read ahead and see – the basis is always taken for granted, even if you claim otherwise.

None of the reasons for our failure to understand ourselves stems from a lack of knowledge or intelligence. It is all about psychology. When the mind is explained, these reasons are also explained, as shown briefly herein.

Activeness

We believe in activeness regarding ourselves and others, which means that we act according to our free will. This is a violation of the law of cause and effect. Since causality is the only way to explain anything, this violation alone may account for our inability to explain the mind. Naturally, it relates to the framework ("our religion") presented above. There is no space to detail that here but you can easily see that the "will" is just a part of the framework, and its "free" aspect is just a feeling.

Please note the absurd. In the first place, physics was separated from the discussion about the mind by denying the causality regarding ourselves (the belief in activeness). In the second place, science was trying to connect between the brain and the mind, but it is already blocked.

"I" am the only object in the world that is "exempted" from causality. This is the meaning of activeness – the common perception that "I am doing", which by definition is inexplicable. Also, the entity "I" is not substantial and constantly changing. There is no way to explain how "I" write, yet, showing how a hierarchical structure of frameworks creates the writing (and the perception of activeness) is straightforward.

Objectivity

Reality certainly seems objective and real, and therefore, it is easy to repress and ignore the "little" contradictions that show otherwise, particularly if this serves our interests. However, when seriously discussing the mind, it is impossible to ignore the law of causality. If perception forms due to causes, the familiar results (our seeing) must be biased accordingly. Objectivity itself is a major bias, which also denies the possibility of biases. It creates unrealistic separations which seem real, while the true explanation is based on exposing the overlaps. For example, everything that you see contains a lot of meaning that arises from memory, while it all seems as external and objective. This also hinders the scientific research. There is much more to say in this regard and is detailed in my books.

Conformity, Authority, and Interests

Our tendency to conform and seek acceptance from others might be the most important finding of psychology, presented from various perspectives and as various psychologies. It is merely the major function of the frameworks, as stated above, and it affects also the scientific research. Some things shouldn't be said and they wouldn't be said, even if they are truthful and lead to the sought solution. The objective pretense creates illusions, yet the research is directed by the prevailing ideas, as history proves [see T. Kuhn]. Objectivity itself is such a misdirecting idea. Scientists, like everyone, need to be accepted and take care of themselves. They need to promote their research topics, and it should be "theirs" and not just some truth.

Integration vs. Segregation

By definition, any explanation is integration, and in this case in particular, due to the apparent differences between the mind and the brain. This explanation requires a few steps that seem quite different and may suit different personalities and expertise. Such a long-distance run, with multiple technologies, fits the engineering approach much more than science. In engineering, it is a supervised group work, with defined steps, designed based on painful experience that no one wishes to repeat. Furthermore, unlike science, a solution must be given at a due time. In contrast, each scientist is on his own, and the expected end-result is a paper, preferably such that enables more papers.

Language/Biases

All the reasons presented above clearly overlap and relate directly to the essence of the frameworks. Now I wish to relate briefly to a more detailed level. We use language and regard it as obvious. This is our means for communication in daily life, work, research, ..., also now. The language is part of thought and they are inseparable. The thought uses elements of language and could not occur without them. These words and grammatical rules (language), which we regard as self-evident, are representations that form in the brain as a manifestation of neural interactions. Not only that it is amazing, unlike the common attitude, but also, it must contain biases. Again, there is no space to detail that here. Yet, it is easy to see that words' meanings extensively overlap while they seem as distinct mental objects.

Self and Other Objects

The same representation method underlies all mental objects, and all of them seem inherently separated although their meanings overlap. This structure permits the attribution of different features to different objects. In this manner, activeness is attributed to the self, as different than any other mental object. Since all things are created in the mind by the same regularity, we have nothing different to compare with. Consequently, pointless attributions may seem like objective reality, and the error confirms itself.

Many cases demonstrate this argument but the general idea is probably clear already. These ideas are simple, yet difficult to digest and usually are rejected on the spot. I am objective, I know, I judge everything that arises and fast to reject whatever doesn't fit my position. Everyone so, and the misunderstanding is retained. Above all, we demand substantial things, yet there aren't such – they are all mind-made.

Science as a Religion

The above discussion related to scientists, as those who were supposed to present the solution. However, scientists work in the context of society and must comply with society's expectations. Thus, science did extremely well about nature and miserably failed to use the same knowledge and competence for explaining ourselves. This is a result of the framework, of course.

The admiration of the common people to science is also a result of the framework (in their minds). In a way, science is forced to play the game and serve as the new religion, though it seems that scientists happily do so. Allegedly this is normal and everyone is are doing the same, except that science has a formally declared goal (of searching the truth), which is violated. The big problem is that the current confusion regarding our nature causes endless troubles and suffering. Thus, we have a good reason to scrutinize the foundations.

Publications

The volumes that clarify this matter are:

Our Religion – A free overview, focusing on the focal question – why the mind wasn't explained long ago? The explanation of the mind and of why it wasn't explained, emerge together. The main point here is that it has nothing to do with a lack of knowledge or intelligence. It is all about psychology, our great misconception about ourselves and the world. When this is clear also psychology is clear and simple. It is a manifestation of the brain's representation method, like everything else.

The Mind as Physics – A short article, presenting briefly the "mechanics" of the mind, devoting one paragraph for each related field.

The Foundations of the Mind – The main book that presents the "physics of the mind". It tries to detail the "mechanics" as much as possible and suitable at this point, uncovering the fractal model based on neuroscience. Yet, most of the discussion is at the mental level, showing how the various aspects of life can be explained (and actually, being generated) by this model. The theory is presented within the context of a lite travel story to gain some freedom and attract more readers, yet, due to the wish to detail it may not fit everyone.

Inverted Reality – This format, by Lin Ward, presents only the main ideas and within the context of a story for kids, with funny figures instead of real people. It is made for readers who are looking for fun and not willing to concentrate and check much. However, the dialogs help to make proper emphasis, and the short presentation makes things clearer. The focus is on the meaning, not on the mechanics.

The Way to Wisdom - The mind was not explained due to our wrong intuition, as explained in the other volumes, and improving our intuition is the Buddhist way. The book details the path in the light of the model of the mind, showing it as simple and logical, and as an integral part of everything else. There are many perspectives, but not many things. Actually, there are no things (no-self), as they all form as manifestations of neural processes. This is about a major bias, and overcoming it is attaining enlightenment.

APPENDIX –
Birds, Psychology and Insight

Like someone in the **midst** of water crying out in thirst
Hakuin, thezensite.com

In this appendix, I wish to present to you the last chapter of a book by Lin Ward, which unfortunately is not in English. It follows a couple of cute birds that nest regularly on her balcony. This permits her to follow the story and not only enjoy watching these lovely birds. It involves a different style than the other chapters here and seemingly deviates from the subject of this book, but it does not. The main point in Lin's book is to show the consistent regularity, that formed by evolution, and therefore is true also regarding ourselves. Her last chapter is simply amazing. A true story, with pictures that supports it. It is a little funny, a little sad, and surly revealing. When you understand the nature of the mind, this story doesn't seem so strange. These birds can teach us something if we care to observe and admit the similarity. They demonstrate our main point in every clearly, presented visually (pics), and thus may facilitate the understanding better than many words.

After losing their chicks, the devoted couple took good care of their new offsprings; they grew, and left the nest, and so did the parents. When they returned, I was glad, to be honest. Even when they stood on the rice pot, preparing to jump inside, I was not angry. Yet, I showed them gently that they do not belong there. It surprised me that these great hunters are interested in my rice. But since they asked for it, I gave them some at the corner of the balcony. They ate it all, repeatedly, over a few days. They liked to take some rice in their beak and fly with it to one of the trees, ...and so it begun.

One day another bird followed the couple to my balcony. She was a little smaller, and position herself on the iron bar, about 40 centimeters above the rice. Her body was shaking and she was tweeting at a constant pace. A closer look revealed that her wings made short and fast moves, which appeared as shaking. Her mother jumped to her from the pile of rice and served the food to the youngster's mouth. It repeated many times; the daughter was standing in her place, while the mother jumping back and forth, feeding her daughter with her mouth. When the mother was busy about other things, the daughter was chasing her, demanding to be fed, while she is shaking and tweeting loudly. She did so even while standing on the rice. She wouldn't eat and beg her parents to feed her. If I wasn't seeing it with my own eyes, I would never believe that such a psychological disorder is possible for birds.

Begging her father.

The begging succeeded

The next day, the amazing couple arrived with two daughters following them. Both stood on the bar, shaking, tweeting, and begging to be fed. Flying to here was easy, chasing their parents aggressively all over the place was not difficult for them, but to pick the rice from the floor by themselves, they wouldn't consider it. This phenomenon repeated a few times, yet, when they were left alone on the balcony, the shaking and tweeting stopped, and they served themselves well.

Happy family

Due to the daughters' attachment, they seemed like a joyful family, a real party on my balcony. Yet, probably this is not the way they saw it since the parents were already preparing the nest for the next generation.

In the first days the mother was taking most of the feeding burden, but then it changed, and the daughters began to chase only the father, bypassing their mother swiftly if she stood on the way.

Both begging

One is served

The other one too

Three days later the mother began to attack them once in a while. Not aggressively, but enough to distance them from the rice and the father for a few seconds. Concurrently, the father was ignoring his begging daughter, that due to having no choice, was begging and eating simultaneously. A reasonably structured maturation process.

One of the young birds understood earlier and left, while the other was still chasing her father. It didn't take too long, and also the father attacked her, but only in the presence of his mate, and without physical touch. These attacks became more consistent each day, where the mother mainly encouraging from the side. She was approaching in a threatening way but as a rule, kept some distance. The miserable daughter was evading the attacks, yet, not running away, trying to get some food, and now completely alone.

Approaching in a threatening way

Father attacking

Youngster evading

Interestingly, the stubborn youngster did not give up and occasionally appeared with her parents. Her begging stopped and their attacks subsided, and they seemed to be in harmony. Everyone learns! and psychology and behavior change accordingly.

The young sisters kept coming, together or alone, just to enjoy a free meal. However, when a parent appeared the shaking resumed even when they were eating. How could they meet in this vast area and coordinate the visit? Something in their memory constitutes the required similarity. The same question is relevant to the parents, even to a greater degree. They cannot speak, often long-distance apart, yet correlate so well.

Birds' psychology, would you believe? I guess that without the pictures no one would. However, what all this mean about ourselves? We are so convinced that we are unique and complex, that only we possess consciousness, while here small birds demonstrate a developmental disorder, similar to those of humans. The poor youngsters acted very inefficiently due to fixation in their brains, which formed a representation in their minds, that seemed very real and vital to them. There was nothing that could change their minds, although the situation seems clear and simple.

Our main subject is not birds, and not even humans, but rather the general regularity that evolved during millions of years and is true to all species, including ourselves. We are the top stage of this evolution, hence, we are different only regarding some details while the principles must be the same. In short, the main theme is the way things are learned and represented in the brain. "Representation" relates to the transformation between the physical processes (brain) and their mental manifestations (perception, recall, and behavior). The simplified model that is presented at the beginning of the book, easily explains the foundations of the mind, the normal functioning, and the disorder presented here. Other psychological aspects, like subconsciousness, repressions, projections, and denials, are also demonstrated as integral parts of the fractal regularity that underlies the mental process, as shown by the model [4]. It means simplicity and uniformity.

This attitude differs strikingly from the normal approach of western psychology, which emphasizes complexity and uniqueness, and uses impressive terms, and as many as possible. This way cannot see the forest for the trees, and very difficult to understand. In contrast, when the focus is on the principles and the essence of things, even "high" terms like self, consciousness, logic, and cognition become simple, and their overlap is apparent. Thoughts and speech are also emerging in the same brain, by the same "method", and are explained similarly.

112

The simple truth, which we all know but refuse to acknowledge, is that all aspects of psychology and the mind are forming in the brain and are subject to its representation method, and it is the same method for all brains, and regarding all their functions. This concerns also the psychological research, of course. It is a mental (psychological) process, and cannot be objective. The state of psychology certainly proves this case, yet in the research, the lack of objectivity is attributed only to the "subjects", never to the researchers. This kind of bias was already mentioned above; it is an unavoidable result of the attribution process in the brain, and it creates the need to constantly observe many perspectives for proper understanding.

The principle of a physical process (as complex as it may be) that creates perception, means that there are inherent biases, which are present all the time, in contrast to the feeling of objectivity that controls us. The basic idea is very simple and the biases were discussed above, yet, they constitute the circular trap that prevents true understanding. Thus, although these topics and facts are not at all new, they are repressed and denied. Allegedly, we wish to know and understand, but it is far more important to us to glorify ourselves and be separated from physics and the world. Like the young birds, we are believing in a reality that does not exist. This reality is maintained due to fixations in our brains, and we ignore the simple facts that belie it. In particular, we are blind to the main direction – this is not an external, objective reality. The Buddhist enlightenment is the release from this fixation. Its purpose is to stop paying the heavy price that stems from the current confusion.

Isn't our state somewhat resembles that of the young birds, that chase their parents while their body shaking, in order to get something which is located under their feet, and does not require any effort

113

to obtain? Similarly, also our need for belonging and connection to the framework generates biases and endless efforts to attain that which was not lacking in the first place. The Buddha's wisdom is the dismantling of the top framework, and then nothing is lacking, not because everything was attained. Simply, the need was not real, and that illusion (the framework) was ended. Unlike the normal belief, having consciousness is not unique to us, but the capability for Buddhahood is, and not attaining it is not fulfilling our lives, although it is not truly open for everyone.

References

1. Suzuki, D. T., *Living By Zen*, Samuel Weiser, York beach, Maine, 1994.

2. Suzuki, D. T., *Zen Buddhism, selected writings of D.T. Suzuki*, Doubleday, Garden City, New York, 1956.

3. Suzuki, Shunryu, *Zen Mind, Beginner's Mind*, Weatherhill, New York, 1995

4. Notmi, U. *The Foundations of the Mind*, 2019

5. Bāhiya Sutta, translated from the Pali by Thanissaro Bhikkhu 1994
https://www.accesstoinsight.org/tipitaka/kn/ud/ud.1.10.than.html

6. Lauren Slater, Opening skinner's box

7. Zen Master Low, Albert (continued by Jean Low) *Thoughts Along the Way*,
https://albertlow.wordpress.com/2015/01/01/hui-nengs-flag-koan-29-from-the-mumonkan/

8. Philodynamics
https://sites.google.com/site/jdquirk/articles/not-one-thing-exists

9. Dong, Vivian and McDaniel, Jay All is
Void https://www.openhorizons.org/all-is-void-and-there-is-no-buddha.html

10. Kuhn, Thomas, *The Structure of Scientific Revolutions*, The University of Chicago Press, Chicago and London, 1962

11. Ussivakul, Archan Vinai, An introduction to Buddhist meditation for results: part one: tranquil meditation; part two: insight meditation Publisher: Bangkok: Tipitaka Study Centre, 1995

12. Henepola, Gunaratana, *The Path of Serenity and Insight: An Explanation of the Buddhist Janas*, South Asia Books, Columbia, 1985.

13. Thera, Nyanaponika, *The Heart of Buddhist Meditation*, The World of the Buddha Publishing Committee, Colombo, 2nd ed, 1956

14. Aran, Lydia, *Buddhism*, Dvir, Tel Aviv, 1993.

15. Pervin, Lawrence A.; John, Oliver P., *Personality: Theory and Research*, John Wiley & Sons, New York, 1997.

16. Goleman, Danial J., "The Buddha on Meditation and States of Consciousness" in Shapiro Deane H.; Walsh, Roger N. eds., *Meditation: Classic And Contemporary Perspectives*, Aldine, New York, 1984.

17. Henepola, Gunaratana, *Beyond Mindfulness in Plain English: An Introductory Guide to Deeper States of Meditation,* Wisdom Publications, 2009.

18. *Vipassana Bhavana*, second edition. Boonkanjanaram Meditation Center, written by Chua Jantrupon and Vitoon Voravises, according to the teaching of Aachan Naeb Mahaniranonda.

19. *WALK TO BE THE KNOWER - Life of the Buddha, What did the Buddha teach, Handbook of Vipassana Meditation*. Wat Tam Wua Forest Monastery.

20. Allport, Gordon W., *The Nature of Prejudice*, Addison-Wesley, Reading, Mass., 1979.

21. Robert E. Buswell, Jr., Donald S. Lopez Jr.*The Biggest Misconception about Buddhism*, Nov 02, 2017, https://tricycle.org/trikedaily/biggest-misconception-about-buddhism.

22. Venerable Phra Dhamma Theerarach Mahamuni, *The Sixteen Stages of Insight*, 1988, Vipassana Dhura Meditation Society, http://www.vipassanadhura.com/sixteen.html